DOING SURVEY RESEARCH

A GUIDE TO QUANTITATIVE METHODS

Third Edition

DOING SURVEY RESEARCH

A GUIDE TO QUANTITATIVE METHODS

THIRD EDITION

Peter M. Nardi

Paradigm Publishers
Boulder • London

Answers to the new "Test Yourself" exercises for each chapter can be found on the publisher's web page for the book: http://www.paradigmpublishers.com/Books/BookDetail.aspx?productID=356915

A website has also been created for this book where readers can find additional guidance, research ideas, references, and examples for each chapter. Post ideas and provide feedback at http://doingsurveyresearch.wordpress.com.

Copyright © 2014 by Paradigm Publishers

Published in the United States by Paradigm Publishers, 5589 Arapahoe Avenue, Boulder, CO 80303 USA.

Paradigm Publishers is the trade name of Birkenkamp & Company, LLC, Dean Birkenkamp, President and Publisher.

Library of Congress Cataloging-in-Publication Data
Nardi, Peter M.
    Doing survey research : a guide to quantitative methods / Peter Nardi.—3rd Edition.
pages cm
    Includes bibliographical references and index.
    ISBN 978-1-61205-306-6 (pbk. : alk. paper)
    1. Social surveys.   2. Social sciences—Research—Methodology.   3. Sampling (Statistics)   I. Title.
    HN29.N25 2013
    300.72'3—dc23
                                                                2012049141

Book design by Straight Creek Bookmakers

Printed and bound in the United States of America on acid-free paper that meets the standards of the American National Standard for Permanence of Paper for Printed Library Materials.

18 17 16 15 14   1 2 3 4 5

contents

Understanding how to describe findings using graphs, tables, and statistics is the focus of this chapter. In addition, the emphasis is on learning to make decisions about when to use the mean, median, mode, and standard deviation and understanding the concept of the normal curve and z-scores. Concepts of probability and statistical significance are introduced.

This chapter discusses how to read and construct cross-tables of data and decide which statistics to use to measure association and correlation. Understanding how to reject or accept a hypothesis by using the appropriate statistics to assess bivariate relationships is highlighted.

This chapter shows how to assess differences between means using t-tests and analysis of variance. As with other bivariate data analysis, knowing when to use these statistical procedures and how to interpret them is central to testing hypotheses.

This chapter focuses on the analysis of three or more variables to answer more complex research questions. It discusses when to use and how to interpret multiple regression analyses and how to perform elaboration techniques with control variables.

In this final chapter, learning to write a report of the research project is emphasized, along with the key elements that go into a presentation of your study. Understanding the different audiences reading a report guides the preparation of the findings.

preface to the third edition

Learning to do research, make decisions about which statistics to use, and interpret statistical analyses is a goal for everyone in today's research-oriented world, where questions about the reliability and validity of data from a study or public opinion poll come up routinely. This book helps in achieving those objectives. *Doing Survey Research* is intended for people who want to learn how to conduct quantitative studies for a project in an undergraduate course, a graduate-level dissertation, or a survey that an employer has commissioned. And it is for those who want to understand and critically interpret research found in scholarly journals, reports distributed in the workplace, and survey findings presented online via a blog or website.

FOCUS OF THE BOOK

The core goal of the book is to provide hands-on experience in developing, carrying out, and concluding a quantitative research project, while keeping in mind the ethical dimensions throughout the research process. This is accomplished by organizing the chapters according to the sequence of steps in which most quantitative survey projects are conducted:

- Understanding the purpose of doing research and the strengths of survey designs compared with other methodologies (Chapter 1)
- Reviewing previous research in the professional journals to clarify ideas and develop hypotheses (Chapter 2)
- Learning to formulate and operationalize valid and reliable measures for a set of research questions or hypotheses (Chapter 3)
- Constructing a professional questionnaire by learning the techniques and guidelines used by survey researchers (Chapter 4)
- Designing quality sampling strategies (Chapter 5)
- Learning to use basic statistical techniques and decide which statistics to use with a variety of measures (Chapters 6 to 9)
- Reading and evaluating tables of data and statistics that appear on the Internet and in newspapers, reports, and academic journals (Chapters 6 to 9)
- Interpreting the data and drawing conclusions about the research questions or hypotheses when writing different kinds of reports (Chapter 10)

NEW TO THE THIRD EDITION

- Each chapter has been updated with data and examples from contemporary academic and popular articles relevant to today's Web-oriented students, including ones focused on topics related to social media, websites, and blogs.
- A new section of "Test Yourself" exercises has been added to each chapter with answers provided in the appendix for students to quiz themselves on the chapter's content. Answers are also posted on the publisher's web page for the book: www.paradigmpublishers.com/Books/BookDetail.aspx?productID=356915
- Another exercise asks readers to decide what to do at each step of a project in terms of making and maintaining friendships, a topic of central concern to the "Facebook connected." This exercise appears in each chapter, providing readers continuity between chapters as they learn new concepts, build on previous knowledge, and demonstrate cumulative knowledge.
- Connections among questionnaire design, online surveys, and statistical analysis are strengthened.
- Getting students to learn how to interpret statistics, not just calculate them, is emphasized.
- Examples from research have been updated to reflect a greater diversity of subject fields and topics, such as sociology, education, political science, health, social work, psychology, and communications.
- A *Doing Survey Research* website has been created where readers can find additional guidance, research ideas, references, and examples for each chapter of the book. Post ideas and provide feedback at http://doingsurveyresearch.wordpress.com.
- An updated author-written Instructor's Manual with test questions is available.

EXERCISES

In addition to the newly added "Test Yourself" questions, there are four other kinds of exercises at the end of each chapter.

- "Review: What do these key terms mean?" The reader is asked to review the key concepts and definitions presented in the chapter.
- "Interpret: What do these real examples tell us?" The reader is encouraged to make sense of information and data from research journals, public opinion surveys, and reports.
- "Consult: What could be done?" The reader is given an issue or problem that requires advice about what could be done, based on the new material presented in the chapter.
- "Decide: What do you do next?" This exercise carries over from one chapter to the next and asks the reader to design the next step in a research project.

A few readers have asked for coverage of qualitative methods and more advanced statistics, such as logistic regression. However, this is a *beginner's guide* to quantitative survey methods rather than a comprehensive, encyclopedic volume on all sorts of social science research techniques. The objective is to keep the book brief, simple, and lively. I see it as preparing a foundation for those students who will go on to graduate school, where they will learn more advanced methods and statistics. And I view the book as a how-to manual for those who do not plan to go beyond an introductory course, yet need to learn (or relearn) how to read survey data and make decisions about which statistics and methods to use when designing a study for a thesis, class, or workplace project. I have received many e-mails from nonstudent readers in "real-world" workplaces who have been tasked with writing a questionnaire, conducting a survey, and analyzing data and picked up the book to refresh their memory of material learned in a course taken years before. These are all the audiences for which this book is written.

ACKNOWLEDGMENTS

I continue to acknowledge all the students at Pitzer College who helped me with this book by reading it, asking me to clarify passages, and providing me with feedback in my quantitative research methods course. Thank you to the anonymous reviewers who provided suggestions and to those readers who have e-mailed me with comments. This book is made possible thanks to the production team at Paradigm Publishers—Ashley Moore, Jennifer Kelland, and Jennifer Top; Annie Daniel in college marketing; and the esteemed Paradigm editor Dean Birkenkamp, not only for his expert advice but for his friendship and persistence. Most importantly, Jeff Chernin continues to make it all possible in every way imaginable yet impossible to quantify.

Everything that can be counted does not necessarily count; everything that counts cannot necessarily be counted.

—*Albert Einstein, physicist*

WHY WE DO RESEARCH

Research is formalized curiosity. It is poking and prying with a purpose.

—*Zora Neale Hurston, writer*

LEARNING GOALS

In this chapter, the differences between everyday thinking and scientific thinking are discussed. An argument is made about the advantages of doing survey research and understanding various kinds of research: exploratory, descriptive, explanatory, and evaluation. The chapter concludes with a comparison of quantitative and qualitative research methods. By the conclusion, you should be able to give examples of everyday thinking, discuss the components of scientific reasoning, and describe the different types of research methods.

Remember the time you asked one of your parents to buy a particular pair of expensive running shoes, let you go to a party, or do something just because "all the kids" were? Well, the astute parent would question this and wonder if it were really true that *all* your friends were actually involved. "Why not collect some data first and then report back to me what percentage of your friends own those clothes or are getting their driver's licenses?" That's what parents trained to do survey research might say!

Too often we make conclusions about entire groups of people based on observations of only those we know or see around us. Or to put it more scientifically: Too often we generalize about an entire population based on a nonsystematic method of collecting data from a biased sample. This is one of the major differences between everyday experience and scientific thinking. What follows is an argument for doing things more systematically and scientifically than usual, and a justification for doing quantitative survey research as one approach to understanding how and why humans think and act the way we do.

EVERYDAY THINKING

How we see the world around us is shaped by a variety of forces that include the books we read, the television and movies we see, the culture's rules and guidelines for good behavior we repeatedly hear, and the teachings from the religious organizations and schools we attend. It is also highly influenced by our friends—the peer groups we spend a good deal of time in and from whom we learn. But so many of these social forces are linked to each other in ways that tend to reinforce our already held values and beliefs. Conclusions based on networks of friends and family members are hardly useful, then, in trying to understand how most other people think or behave.

The trick of socialization, to paraphrase writer Carlos Castaneda, is to convince us that the way we see the world is the only reality, one supported by social consensus. By experiencing more diverse cultures and meeting a wider range of people, we come to understand that to make conclusions based simply on the way our friends and family live, what they believe in, or how we were socialized, is limiting. Too often those with whom we associate are similar to ourselves in values and beliefs; we end up selectively sampling like-minded people and erroneously concluding that "everyone" experiences reality in the same way we do.

Although we seem to manage fairly well on a daily basis, our everyday experiences are often based on methods that can lead to problematic decisions with outcomes that can seriously affect our lives. Imagine if some friends told us that they were able to stop flu symptoms and prevent a recurrence of a cold simply by eating only the white filling in a chocolate cookie every day for a month. Would we rush out to buy bags of these wonderful cookies and use them to ward off the flu instead of getting a vaccine? It certainly would be more fun (and fattening) but not likely to keep us from getting the flu. What if they told us to take vitamin C? Would we run to the store to get some, or would we critically inquire about the research that tests this general statement?

Every day we make conclusions and act on them with similarly limited information. Recall how sometimes you get a text message from someone you suspected would be getting in touch: "I had a feeling you were going to send me a message."

Never mind that you don't remember the number of times you did not guess accurately or that you normally contact that person around the same time every day.

Or recall the time you ran to the nearest movie theater to catch a film your friends exclaimed to be the best ever and some critic called "the best film of the year," even though it was only March. Although it would be impossible and crazy to survey a large number of people every time we need to make a decision to do something, be aware of how the process of making a choice is often not much more systematic than believing the person with the secret cure for the common cold. Yet, we make such decisions because we know from experience ("empirical data") that this particular friend's recommendations have almost always been reliable.

What we do in everyday life is typically the result of some less scientific thinking than the procedures we would expect policy makers, neurosurgeons, or airline pilots to follow when they are in control of our well-being. In other words, what we need to consider is how ordinary thinking differs from the systematic methods needed to understand complex social behavior and attitudes. By doing so, we begin our research journey on the correct foot.

The characteristics of ordinary, everyday thinking and inquiry include

- Biased questions
- Limited sampling
- Selective attention, perception, and retention
- Inaccurate generalizations

Biased Questions

Did anyone ever ask, "Do you really want to go to that Lindsay Lohan movie?" and say it with an air of disdain? It would be difficult to answer, "Yes, I really do want to see it" since the questioner was implying that the film was so bad no one in his or her right mind could really want to go. And remember the time someone wanted to know the reasons why everyone was so dissatisfied with the workplace, thereby assuming that "everyone" was unhappy and all we needed to do now was find those reasons? Consider the research question usually worded this way: "Why is it that I study more than the other students yet they get better grades?" and the implications already formed by the phrasing.

In each of these three cases, everyday nonscientific thinking leads us to ask questions that are already *biased* or slanted in a particular direction. The first, perhaps more rhetorical question is already providing the answer in the way it is worded. It does not allow for a full range of possible answers, so the only conclusion would be that everyone who was asked the question in this way in our survey does not want to see the latest Lindsay Lohan movie. We'd be making an accurate statement of what was uncovered, but the answers are only as good as the way the question was framed and applicable just to the *limited* sample. The tendency, of course, is to *generalize* to all people, that is, to make conclusions about people not surveyed using the information obtained from those that were.

Similarly, the second question proceeds from a *selective* assumption that may not be accurate, so the only answers we get from a survey of employees' experiences are negatively *biased* ones about the organization. How many times do we really ask, "Why is this workplace such a happy one?" with the same enthusiasm and inquisitiveness as we asked why people are dissatisfied with work? Or, more accurately, how often do we ask what they think about their workplace, without qualifying it as good or awful when we inquire?

With the third question about study habits, we are talking ourselves into believing something that may not be true, but asking it in this form just the same. Rarely do we, in everyday talk or thought, put it the following way: "I wonder if there is a relationship between the amount of time people study and their grades." We often begin with a particular viewpoint and proceed to ask questions from that selective position, sometimes not even aware that we are doing so. Thus, the answers we get

BOX 1.1
EVERYDAY BIASES

A great source for studying how people selectively see the world, make use of limited samples, generalize to people not studied, and ask questions in biased ways is to read the "Letters to the Editor" in newspapers or the comments section of online blogs and publications. Soon after a major shooting occurred in a suburban high school, a series of comments appeared in the paper's online story, each with its own explanation for what happened and each likely representing the writer's own personal values and biases. These comments included the following reasons:

1. Life in white monotonous suburbia and the alienation that results
2. Republican congressmen who refuse to pass legislation controlling guns
3. Parents who neglect their kids' depression and isolation
4. Schools that allow hate-filled speech and all sorts of taunting and teasing of kids who are different
5. Legal abortion and the theory of evolution, both of which lead to a devaluation of life

Take a look at your local paper or the comments for an online magazine column or blog and see how many examples you can find of everyday, nonscientific reasoning.

tend to verify what we already believe and give us a false sense of having found some objective and honest answers. We now have the reasons why our *limited* sample of employees hate the cafeteria food and falsely *generalize* that everyone hates the food because we never found out how many liked it in the first place.

Our daily interactions with friends, family members, and the people we encounter on a routine basis rarely require anything more scientific. Ordinary discourse contains many such poorly worded questions and biased assumptions, and the world manages to keep going. However, imagine the consequences of asking questions in these ways when trying to understand more complex and important behaviors and opinions or when attempting to develop public policies that can seriously affect people's lives. No one should think that an educational institution, for example, can create rules to require a certain amount of study hours per week simply based on the findings of a survey that asks only why those who study less get better grades. Should the human resources office really modify an organization based on the results of a survey focused solely on the dissatisfaction of its employees?

Everyday thinking, in other words, typically employs biased phrases and other nonscientific styles of framing questions when making sense of the world we live in. The results of these queries may help us decide which movie to watch, but they are not useful when it comes to making choices that affect social policies or arriving at conclusions fairly and consistently. How we can improve the way we ask questions and minimize the biases that derive from faulty wording is discussed in more detail in Chapter 4.

Limited Sampling

Most of the time, we make sense of reality by reflecting on the experiences we've had. It is not too difficult to figure out how our friends feel about various rules, political leaders, music, and television programs. We are generally good at assessing the climate of opinion about controversial topics among our peers and those whom we encounter regularly in our living spaces. Unfortunately, there is also the tendency to take these limited experiences and then assign them to larger groups of people.

"Everyone I know hates the food in the cafeteria"—so it must be bad. "I talked with others in the class who have good grades and they're not reading the assignments"—so it doesn't matter how much one studies. "No one I know liked that movie"—so it must be failing at the box office. While it may sound reasonable to make such concluding statements, the problem, of course, is that we have simply taken the opinions or measured the behaviors of those we already know and then made the assumption that they are somehow representative of "all" people: "But,

Mom, *everyone* is going to be driving to school this year." But if they are our friends, they are likely to be people who share our tastes and values.

Selective Perception

People forget that most of the time our lives are constrained and limited by the social spaces we inhabit. Race, social class, gender, sexual orientation, and religion are just some of the many characteristics that provide different experiences and push us into living in unique subcultures and communities. Everyday life is often a series of encounters in limited areas. We rarely get to move outside our circumscribed environment and, when we do, we do not always see what is there. It is part of human existence to *selectively* attend to, perceive, and retain information. After we have arrived at some belief about people or assumption about some events, we use these beliefs to focus our attention mostly on those aspects that already fit our assumptions. We typically see only those dimensions, and we tend to remember better just the events that verify our already held beliefs. That's what the process of selective perception, attention, and retention is. It's difficult to experience reality differently from the way we are taught, and contradictory evidence usually shakes us to the core.

How many of us actually noticed what our boss or favorite professor wore yesterday? Most of us don't *attend* to or pay attention to such things unless we have a particular interest in evaluating someone else's clothing styles. And even if we did notice, what aspects of the clothing did we even see? We tend to *perceive* only some things, in this case, perhaps just the shoes or the type of shirt. And when we do make note of them, for how many days do we *retain* this information? Can we actually recall what we had for dinner or what our best friend wore a week ago?

As a result, for us to suggest that we understand how other people feel or behave based on our own limited sampling, selective perceptions, and selective retention is problematic. At best, we could say that we have a sense of the way the people in our lives approach the world. To use that everyday information, though, to enact policies or change the rules of the game for everyone else would be overgeneralizing and unethical. *Overgeneralization* occurs when we attribute patterns to an entire group and make conclusions about a wide range of people or events based on a few observations. The limitations imposed by our everyday investigation of only the people around us restrict our use of that information for scientific purposes. How best to sample and generalize and how to go beyond the convenience of talking to those whom we already know are the topics of Chapter 5.

SCIENTIFIC THINKING

In order to be able to make accurate and reliable conclusions about human behavior, we are required to go beyond the components of everyday thinking. While those techniques and styles may help us get along in our daily lives, the job of measuring various aspects of the social world and people's beliefs and behaviors demands a more deliberate approach. Certainly, intuition can play some role in arriving at conclusions and making decisions, but would we really want someone about to operate on our body to do so guided by today's horoscope or by some hunch and intuition?

We would hope surgeons base their decisions on techniques and knowledge gained from decades of scientific studies and research. Although understanding study habits and grades is hardly as important as brain surgery, if we were to enact some policies and procedures that can have a positive or negative impact on students' well-being, then investigating patterns of studying also deserves some rigorous and systematic methods. Developing policy or drawing conclusions about social life is best accomplished with the assistance of scientific procedures rather than with a dependence on the everyday thinking described previously. Science is certainly not the only way of arriving at information, but it may be better suited for specific kinds of questions.

Scientific thinking is characterized by

- Empirical observations or data
- Systematic and deliberate methods
- Objective, intersubjective, and replicable procedures

Research Design and Empirical Observations

Let's consider what these mean and see how they differ from everyday thinking. In order to be more confident that the findings on which our policies will be based or the conclusions we make about social relationships are accurate, we need to develop scientific methods for gathering our observations. We need a *research design,* or a plan for translating our research objectives into measurable and valid information. Simply making decisions based on what we see happening among our friends and the people we encounter is not a reliable procedure for generating ideas about larger groups or categories of people. Yet, one of the basic principles of science is involved in everyday thinking; that is, we regularly make observations and collect data.

Empiricism states that the primary source of knowledge is experience, especially that gained through the senses. We understand the world by observation (data collecting), not just via speculative thinking or theories. At some point, to be scientific, we must encounter the reality that is out there and experience through observation whether the educated hunches or ideas we proposed in our theories are substantiated.

Systematic and Deliberate Methods

Empiricism is also part of the everyday procedures we sometimes erroneously use to make conclusions, so other elements are essential. These observations must also be *systematic, objective,* and *replicable.* This is where issues of representative and random sampling, measurement reliability and validity, and other methods used to make observations come into play, as discussed in later chapters. By spelling out clearly the details of how we are to measure and whom we are to observe—and "what" we will do to get the "who" to do the "how"—we are engaging in the methodical step-by-step procedures of research design that make scientific thinking more *systematic* and deliberate than everyday thinking. Although such procedures don't always guarantee completely accurate results, they do eliminate many of the errors that are part of ordinary, nonscientific observation procedures and allow us to generalize and arrive at conclusions about larger numbers of people or events.

Objective, Intersubjective, and Replicable Procedures

Some might argue that the elimination of errors and biases is what makes science more *objective,* that is, less dependent on emotion or personal prejudices and values. Because the procedures have been systematically detailed, other researchers could *replicate* the study (i.e., repeat it using the same methods with a new sample) without interference of individual biases. Of course, researchers have biases and hold a wide range of values that not only may affect the topic they choose to study, but could also influence the procedures they develop to make the observations and collect the data. After all, scientists engage in everyday thinking and employ selective perception as well. This is why replication is a necessary step in the scientific process.

There is nothing wrong with this subjectivity; indeed, it is a key process in uncovering the multiple ways people understand reality differently. Rather than assume that certain words or concepts are objective, we sometimes need to recognize that there are alternative meanings attached to those concepts. For example, when a questionnaire inquires about marital status and the range includes "married," "divorced," "widowed," and "single/never married" categories, how are these options viewed by gay and lesbian respondents who are living with someone of the same sex in a romantic relationship? What if respondents were cohabiting with and not legally married to an opposite sex partner for 15 years and the relationship ends, are they still "never married" or "separated/divorced?" Subjectivity provides the insights and meanings others have about social behavior and attitudes and contributes new ways of developing measurements more comprehensively. We cannot, however, use subjective perspectives as a sole research method if we plan to make scientific, generalizable observations.

If the procedures are comprehensively described, and other researchers with differing values and beliefs replicate the methods in their studies and achieve similar results, we can more confidently conclude that the methods and findings are less affected by any personal biases of the researchers. Perhaps a better word than *objectivity* to describe what is going on is *intersubjectivity,* or what Ira Reiss (1993: 6) suggests happens when people with differing perspectives collectively agree on a particular way of seeing reality: "So we can define objectivity in science as those views of the world that come to be agreed upon by the community at any one point in time."

Finding the "truth" about human behavior and social processes is an ongoing goal of the social sciences. Knowledge is achieved and "facts" are built incrementally over time. The information we get at any one point is an approximation of the truth and is constantly being modified, clarified, and expanded with each new study. This is why it is so important to develop scientific ways of measuring and thinking about patterns of social behavior and attitudes in order to achieve the basic goals of research.

THE PURPOSES OF SCIENTIFIC RESEARCH

Research on human social behavior and attitudes is conducted for many reasons, including to explore, describe, explain, and evaluate for the purpose of understanding an issue in depth, arriving at decisions, and making predictions.

Exploratory Research

Sometimes research is conducted for exploratory reasons, that is, to get a rough sense of what is happening on a particular topic for which we don't yet have enough information. People use *exploratory research* to assess the opportunities for undertaking a study, to try out various methods for collecting information for a proposed larger study later on, or to learn the language and concepts used by those who will be studied. Exploratory research also has been designed to ascertain the needs and goals of a particular organization (often called *needs assessment research*) in preparation for a study or evaluation. *Focus groups*—a collection of respondents organized in a group discussion format to present their ideas about a subject—are frequently designed to achieve many of these exploratory objectives.

Let's say we are interested in understanding more about what concerns students about their educational institution that may cause them to drop out or stay in school. Although we should review the many studies that have been done already on dropouts, we have reason to believe that the unique dimensions of this school are particularly important to consider. So we design a study to explore what is going on at this school by developing a series of focus group discussions with students to learn their language, or jargon, uncover relevant topics, and understand different ways of viewing the institution in order to construct a better questionnaire for a later research project.

BOX 1.2
DESCRIPTIVE RESEARCH AND SCIENTIFIC THINKING

If we were interested in understanding drug use and nonuse patterns among Asian American teenagers and depended on everyday thinking, we could make conclusions merely by looking around a high school and arriving at some description of who was using what and how much. Yet, this "research" would in no way be systematic, and we might erroneously make inaccurate generalizations. Consider instead this excerpt from a study by Nagasawa, Qian, and Wong (2000: 596), which uses data from the Asian Student Drug Survey funded by the National Institute of Drug Abuse. Notice how a more scientific data–based approach results in descriptions that remove some of the errors of everyday thinking and false generalizations to an entire racial/ethnic category:

It is inappropriate to lump Asian and Pacific Americans together into a single racial/ethnic category. Our data suggest that Asian and Pacific Islanders differ significantly in drug and alcohol use. For example, Pacific Islanders have the highest use of marijuana and cocaine. Among Asian Americans, Filipino youths are most likely to use marijuana and cocaine, while Chinese youths are least likely to do so.

Descriptive Research

A typical goal of exploratory and almost all kinds of research is to provide basic information describing the topic and respondents involved. *Descriptive research* is often the first step in most research projects and the primary objective for some, like the U.S. Census or the General Social Survey (GSS) and similar large surveys designed for gathering information. If we are interested, for example, in understanding the relationship between the number of hours spent studying and grades earned, we need to get descriptive information about the characteristics of the students (gender, race, major, etc.), what their grades are, how many hours a week they study, how many years of education they already have, and how many hours a week they work or participate in sports or other activities. The goal of such a survey would simply be to present basic information profiling the respondents (referred to as the *demographics*) and describing the issues under study.

Explanatory Research

Once we have some descriptive information, we might then want to uncover the reasons why a relationship between grades and hours spent studying differs for the students we sample. We do this kind of research in order to *explain* relationships, to uncover the reasons "why" or "how" some social phenomena occur among respondents. *Explanatory research* is designed to answer the "why" question: why there is a range of behaviors or opinions held among people surveyed. Why don't they all have the same attitudes toward capital punishment or vote in the same way? Ideally, we want to be able to explain or perhaps predict these opinions and behaviors with efficient, less complex reasons or causes.

Remember when a romantic relationship ended and we wanted *the* explanation or cause for its not working out? We'd all like to have just one simple reason (such as "it was the other person's fault"), but we know down deep that there are many reasons, since most complex behaviors and attitudes require more than one cause. However, the "law of parsimony" suggests that we should look for the fewest number of reasons (causes, explanations) to account for most of the differences (variation) that exist among our respondents in the behaviors or attitudes we are attempting to study.

Evaluation Research

We are typically interested in understanding the causes of human social behavior and people's opinions about a variety of issues. Sometimes, however, research is conducted to evaluate specific outcomes and to provide the explanations for why and how a particular result occurred. Applied or *evaluation research,* as it is called (see Rossi et al. 2004), focuses on problem solving and measuring the results and specified outcomes of the implementation of various social programs and policies (the "causes"). Many educational institutions also develop evaluation tools to assess students' achievement of the schools' intended objectives and goals.

For example, a university wants to know whether its study abroad program results in students achieving an intercultural and international understanding of people and issues. It creates a new type of program designed to produce such understanding. Evaluation research would focus on the objectives of the program and assess whether the new study abroad experience directly resulted in any change in students' behavior and attitudes.

Decide and Predict

With information collected systematically, those responsible for a program or policy can make informed decisions about what dimensions need to be changed, enhanced, or removed. Using research to *decide* and *predict* outcomes is a central goal of much research, especially evaluation and explanatory research. By figuring out the causes of behavior or opinion, we use this information to make informed decisions about future events. One goal of research is to estimate what might happen after the research is completed, that is, to make forecasts about a company's future earnings, or to estimate the impact certain social policies will have or to guide us in making *decisions*. Wouldn't it be great if teachers could present information to their students demonstrating that there would be a high likelihood of success in school if they put more hours into their studying? Wouldn't it be a worthwhile goal of research to be able to understand the causes of earthquakes or tornadoes and use this knowledge to *predict* their occurrences and make decisions about building codes and other modes of disaster preparation?

Cause and Effect

However, to determine causation, it is essential to (1) establish that a relationship (or correlation) exists between the alleged cause and the observed outcome or effect; (2) determine the timeline of occurring events, that is, the cause must precede the behavior or opinion in time; and (3) eliminate other plausible explanations or alternative causes. If we can show that (1) there is a connection between hours of studying and grades; (2) the grades come after the amount of study hours claimed (it is possible that someone with low grades could increase the number of hours studied as a result of the grades); and (3) no other possible explanation exists for the grades, such as number of hours working at a job, teaching quality, or time spent at parties, then we can more confidently conclude that how much students study explains a good deal of why they achieve various grades. We usually are not able to explain why one particular person does well or not; instead, we may have found an explanation for variation among a sample of students. Most social science research is focused on understanding differences among aggregate groups of people, not in explaining or predicting one individual's behavior or opinions.

All three of these elements must be present to declare a cause and effect relationship. Too often, in everyday thinking, people assume that correlation is the same as causation. Just because two variables are related does not mean we can conclude that one *caused* the other.

Many humorous examples of false causation based on correlation have been offered by researchers. A good illustration is a correlation between the number of fire engines at the scene of a fire and the amount of damage done. Although a strong relationship can be demonstrated between more fire engines and worse damage, obviously the fire engines don't cause the damage. Similarly, just because most heroin addicts drank milk as children does not mean that milk is a gateway substance leading to heroin use and causing addiction. (See Box 1.3 for another example.)

Despite being aware of the clearly faulty reasoning in jumping from correlation to causation, researchers can make this logical error in their interpretation of data because it is often impossible to control for all available alternative explanations. Many studies, for example, show a relationship between high-fiber, low-fat diets and lowered risk of heart disease. Does this mean that eating such foods causes lower rates, or could some other explanations be involved? Some researchers have found that the type of people who eat healthy foods also tend to smoke less, exercise more, and experience less stress. Could these behaviors, instead of diet, be the more direct cause of healthier hearts? Only with replication of studies demonstrating continuing strong correlations, controlling for these plausible explanatory variables, and clearly delineating the timeline of which behaviors come before others can researchers conclude with confidence that certain foods lead to better health.

Concluding Thoughts on Scientific Methods

For us to describe, explain, and predict with any accuracy, it is necessary that we develop a research design of scientific procedures and avoid the kinds of everyday thinking that could lead to incomplete data and erroneous conclusions. While it is often tempting to make decisions, explain causation, and predict the future based on psychic powers, intuition, and "gut feelings," these are not particularly useful skills for convincing funding agencies, public policy makers, or research methods professors that what we have uncovered is accurate and unbiased.

This is not to say that scientific procedures are always ideal and problem-free. There are many behaviors and opinions that elude the methods of science. How well can we really measure and explain the process of falling in love, predict who

BOX 1.3
CORRELATION OR CAUSATION?
STORKS, VACCINES, AND CAUSATION

By Peter M. Nardi

Before learning about the "birds and the bees" we may have been told how the stork brought us, as a little baby, to our parents. Even with a minimal interest in the animal kingdom of storks, birds, and bees, we likely started to question this curious story.

That is until we heard this news about Denmark: Post-1960 there was a significant decline in the number of nesting storks in Denmark (Dybbro 1972). Also, beginning in the late 1960s, Denmark started recording its lowest average number of childbirths per woman. In short: fewer storks = fewer babies.

Here rests one of the fundamental errors in debates, research, and uncritical thinking: confusing correlation with causation. So powerful are spurious relationships that they can sometimes have significant public policy implications. Consider the story of autism and vaccines.

In 1998, *The Lancet,* a respected medical journal, published Dr. Andrew Wakefield's research claiming a link between autism and the MMR (measles, mumps, rubella) vaccine. Ever since, people in the autism community have raised concerns about live-virus vaccines and their children's health. Fueled by the popular media, in particular the Internet, Dr. Wakefield's research has resulted in a decline in vaccinations and, some say, a resulting increase in childhood diseases like measles.

However, on January 28, 2010, Britain's General Medical Council concluded that Dr. Wakefield acted dishonestly, unethically, and irresponsibly when carrying out his research. And on February 2, *The Lancet* said, "we fully retract this paper from the published record" (Park 2010).

Although it's reasonable to have some concerns about the many ingredients that go into vaccines and other medications, it's still important to look more closely at the specific issues raised by the MMR vaccine and autism research, and use our critical-thinking skills in understanding what is going on. When assessing research, it's important to evaluate several elements: the sample, the quality of the data collection process (such as survey item wording or interview style), and how the data are analyzed (appropriate statistics and charts).

Let's begin with the sample: The original Wakefield study took blood samples from only 12 English children who were attending his son's birthday party. They were each paid the equivalent of around $8. Already, we begin to question the quality of the research when such a small sample is used. It's also important at this point to consider any ethical questions about paying the children studied and how they may have been affected by having invasive blood samples taken.

For research to carry any weight, replication is essential, and studies with larger and better samples have not demonstrated a correlation between vaccines and autism. Furthermore, before cause and effect can actually be declared from a correlation, a timeline must demonstrate that the cause came before the effect. For example, students who study more tend to have higher grades. But does studying lead to higher grades, or do those students who have higher grades (maybe who are smarter to begin with) tend to study more to ensure continuation of a high GPA?

BOX 1.3 CONTINUED

When reviewing how researchers collected the data, assessing which data occurred when is important. In many cases, it turns out that autism appeared before the vaccinations were administered.

In analyzing the data collected, in order to claim a cause and effect, review how the research eliminated alternative explanations. Do changes in industrialization and urbanization in Denmark, for example, connect to a decline in the stork population as well as to changes in family life and fertility? Spurious correlations are easily addressed by searching for a third explanation. The appearance of autism tends to occur between the ages of 2 and 5, the same period when vaccines are administered.

Just because there is a societal increase in autism rates coinciding with an increase in the distribution of vaccines, it does not indicate a cause and effect relationship, especially if autism rates continued after thimerosal (the mercury-based preservative hypothesized to be at fault) was removed from vaccines.

Increases in autism rates could be due to other explanations such as changing definitions of autism and better diagnosing techniques, thus illustrating how other variables can create the illusion of a correlation between immunizations and autism. Other studies also indicate that boys are about four times as likely to have autism despite similar rates of vaccination.

Finally, a major study in 2002 of almost half a million Danish children found no difference in immunization records between those children with and without autism. To date, there is no scientific evidence in the published literature of a causal connection between immunization vaccines and autism. And thanks to Denmark we have the research on this spurious relationship between autism and the MMR vaccine—and, of course, on storks and childbirth.

Reprinted from *Pacific Standard* online magazine, March 1, 2010, www.psmag.com/science-environment/storks-vaccines-and-causation-10195.

will successfully be our best friend, or fully understand religious fervor? Many have tried, but sometimes more abstract ideas require other kinds of methods besides quantitative scientific ones—or should be left to the work of poets and artists!

The research problem or the evaluation questions must determine the methods. And for most of the issues social scientists study and the complex behaviors we want to understand, scientific thinking and procedures work well. Yet the method should not precede the problem to be studied. Before we choose a questionnaire survey approach, we should consider the different kinds of methods that can be used to study human social behavior and then select the techniques that best fit the questions we are seeking to answer.

RESEARCH METHODS

Doing survey research is a skill, an art, and an intellectual process involving collaboration, patience, and creativity. As Laumann et al. (1994: 57) claim, "In practice, survey research methods, like many specific scientific laboratory techniques, remain more an art than a science." Survey research is also a choice of one method among many from which to select. As such, choosing to conduct a quantitative approach to understanding the social world can answer only some questions. It is not ideal to begin by saying, "I want to give a questionnaire out, but I am not sure what my topic is yet." Research questions must come first, and then the choice of the relevant method to study them should follow.

There are many different ways of gathering data, depending on the questions we are asking, whom or what we are studying, the financial and time limitations of our project, and the amount of detail we desire. Each method not only comes with strengths and weaknesses that must be evaluated carefully before selecting but also comes with a set of assumptions about the nature of knowledge, beliefs in the efficacy of science, and other philosophical questions about how we can make sense of the world in which we live. Most methods can be combined to study a topic (often termed *triangulation* or *mixed-methods research* when two or more measures or methods are used), and several of them share similar procedures, sampling strategies, and ethical considerations. Larger research textbooks (such as Creswell and Clark 2011 or Babbie 2010) provide more details about the different methods and the scientific assumptions that go with them. Because the focus of this book is on quantitative survey methods, here is a brief overview of other research methods, with key points to consider when evaluating whether questionnaire survey methods are best suited for a particular study. (See Box 1.4 for the advantages and disadvantages of each major type of research methodology.)

Experimental Designs

When interested in understanding how the manipulation of a variable can explain specific outcomes on another variable, some researchers find it useful to conduct experiments (see Campbell and Stanley 1963). A classic *experimental* research design typically involves comparing two groups, one called the experimental group, the other the control group, to both of which respondents have been randomly assigned. In the experimental group, the researcher conducts some treatment on the subjects and measures its effects in comparison to another group that does not receive the treatment or to a group receiving a different kind of treatment.

Experiments typically occur in laboratory settings where the researcher can control the environment to prevent other plausible causes from affecting the outcome of the treatment or experiment, thereby ensuring internal validity or accuracy and perhaps allowing for generalizability, or what is sometimes called external validity (see Chapter 3 for a discussion of validity). Experimental designs are also suited for testing specific hypotheses and for doing applied evaluation research, rather than for conducting an exploratory study.

Imagine we are interested in understanding the impact on teenagers of an educational film about prejudice and racism. In the classic experimental design, we would randomly assign the teens to two groups. Each group completes a questionnaire focused on attitudes toward racial minorities and other indicators of prejudice. The experimental group views the video (the treatment) while the control group does not, or perhaps sees a different one if we were interested in comparing films. A few weeks later, a questionnaire on racism is given to the two groups and comparisons are made. Ideally, those who watched the film now have lower prejudice scores than the group who did not see the film; we infer that the film was partly responsible for this change. There are many variations of this classic experimental model, many of which are used in evaluation research and by experimental psychologists and social psychologists.

Qualitative Methods

If the goal is to understand human behavior in its natural setting and from the viewpoint of those involved, then an appropriate method is often a qualitative one, as opposed to a quantitative method in which predetermined categories and a more structured scientific approach are involved. Qualitative research explores new topics by getting into the settings where people carry out their lives. Anthropologists typically use qualitative methods to understand a culture, and some of the earliest sociologists (often referred to as the Chicago School of sociology) were pioneers in using these methods to study how people lived in small towns and urban centers (Plummer 2001).

Field research, participant observations, ethnographies, case studies, open-ended interviews, and focus groups are some common types of qualitative research methods (see Esterberg 2001). At some level, they all involve observing what people do, what they produce, and how they interact verbally and nonverbally. For example, if we wanted to understand how people make decisions about what food to eat in the employee cafeteria, we might do better observing them than to ask such questions days later on a questionnaire. Going to the cafeteria, taking extensive notes about the kinds of people who choose different foods, observing how much they eat, listening to how they interact with other diners, and talking with them about their choices are just some of the methods of a qualitative approach.

Trying to understand with more depth and sensitivity people's subjective understandings while acting in their social situations is the main goal for qualitative research. We typically do not get to study a very large number of people when using qualitative techniques compared with survey research, but we usually get richer details and a stronger sense of the variety of ways people engage with the world around them. It is a technique ideally suited for doing exploratory research as well.

Content Analysis and Archival Research

It is not always necessary to study people and their behaviors and opinions. Occasionally, we might be interested in understanding what they produce and to see how this might change over time. *Content analysis* involves the study of artifacts, usually written (such as diaries, newspapers, blogs, biographies, Twitter messages, and official documents) but also visual and other forms of communication. It is based on developing a way of coding and classifying the information (the content) in the documents or media being studied. Content analysis is often used in coding answers to open-ended items on questionnaires (see Chapter 6 for a discussion about this). It includes qualitative methods and sometimes the quantification of information. For example, we can study the content of radio talk shows and code the broadcast in terms of how liberal or conservative the views of the host and the callers are, as well as tabulate the percentage of men and women who air their opinions. Or we might be interested in researching the images of gays and lesbians in television shows and do both a quantitative analysis of the number of characters and a content analysis of how they are depicted in these shows: Are they portrayed in stereotypical ways? What are their issues? Are they shown in relationships or isolated?

Let's say we are interested in evaluating online blogs about global climate change. We would develop a sampling scheme, read the postings and construct coding categories, and then evaluate the blogs in terms of those codes, perhaps looking for biases in language when discussing scientific studies and what positions the blog writers take about global warming issues. Sometimes this work involves searching archives and other *historical* documents for information about environmental concerns in newspaper editorials from previous pre-Internet decades, for example. *Comparative* research could also include reviewing blogs and climate change laws in different countries.

Quantitative Methods

Many times researchers are interested in describing the number of people involved in certain behaviors or holding specific beliefs. Some want to make use of archival data that have been collected by others over the years, such as all the information gathered during a census. Others like to focus on explaining the way people behave or predicting how they might act in the future. Underlying all these is an assumption that social phenomena can be systematically measured and scientifically assessed. For many of these kinds of questions and assumptions, the use of quantitative methods is most appropriate, as we have been discussing in this book.

Some of the techniques involved in content analysis, experimental designs, archival analysis, and in-depth interviewing use quantitative approaches. Structuring questions for an interview, developing categories and variables for coding printed content, and counting responses and observations are just some of these techniques. An easy way to remember the differences between quantitative and qualitative research is to think about how someone reviews a movie. A reviewer who goes into details about the acting, the camera work, the screenplay and dialogue, and the grand meaning of the story is providing a qualitative content analysis. A reviewer who simply says it's worth three stars or two thumbs up has given a quantitative response with fewer details but a convenient summary evaluation.

Quantitative methods typically involve writing questions for surveys and in-depth interviews, learning to quantify or count responses, and statistically (mathematically) analyzing archival, historical, or our own data. A common form is a self-administered questionnaire. Questionnaires are particularly suited for respondents who can read, for measuring people's attitudes and opinions, and for getting a very large number of respondents too difficult and time consuming to observe with qualitative methods.

Doing survey research well is the theme of the rest of this book. It focuses primarily on questionnaires with samples of people, but many of the techniques described apply to other situations and methods. The chapters of this book are arranged in the order typically used for generating a research design and writing up the results for presentation:

1. Find a topic to study.
2. Review the previous literature and research.
3. Develop research questions and hypotheses.
4. Specify how to measure (operationalize) the variables in your hypotheses.
5. Design a questionnaire.
6. Develop a sample.
7. Collect data.
8. Prepare a codebook and data file.
9. Enter survey results in the data file.
10. Analyze data statistically.
11. Write up and present the results and conclusions.

The first stop on the research journey is learning to create good research questions and generate problems to evaluate. In the next chapter, we look at ways to find ideas for using the scientific methods of quantitative inquiry.

BOX 1.4
COMPARING METHODS

Each method for collecting data has advantages and disadvantages that should be evaluated before you decide which ones are most suitable for a particular research topic. Here are some points to consider for collecting data with surveys, interviews, focus groups, qualitative methods, and experiments.

METHOD	ADVANTAGES	DISADVANTAGES
Quantitative: Surveys	• Less costly to reach larger samples • Standardized questions • Ideal for asking about opinions and attitudes • Less labor intensive to collect data or train researchers • Can guarantee anonymity • Suitable for probability sampling and more accurate generalizability • Easier to code closed-ended items • Respondents can answer at own pace • Better for sensitive and personal topics • Easier to replicate a study • Can address multiple topics in one survey • Ideal for computer-based and Web-based surveys • Easier to compare with other studies using similar questions	• Self-report requires reading ability in the language (age, eyesight limitations, education) • Possible gap between what people report they do and what they actually do • Return rate can be low for mailed and computer-based surveys, thus limiting generalizability • Closed-ended questions can be restrictive and culturally sensitive or dependent • Difficult to explain meaning of items and probe answers • Depend on asking about recollected behavior • More difficult to code open-ended responses • Can't guarantee respondent answering it was the person intended to answer it • Requires skill in questionnaire design • Long and complicated surveys can be tiring to complete and lead to errors • Easy to overlook, skip around, and misunderstand questions • More difficult to generate reliability and validity for one-time-use questionnaires
Interviews: Structured face-to-face or telephone	• Standardized questions for structured interviews • Can explore and probe for additional information • Can clarify meaning of questions • Telephone interviews are less costly and can reach larger samples • Less likely to have skipped or missed questions • Unanticipated answers can occur, thus leading to new, unexpected findings	• Limited to smaller samples • Face-to-face interviews can be time consuming • Training required for interviewers • More difficult to code open-ended responses and unstructured interviews • Interviewer characteristics (race, sex, age) and style could bias responses • Some respondents reluctant to give information over the telephone • Not as ideal for collecting sensitive or personal information

BOX 1.4 CONTINUED

METHOD	ADVANTAGES	DISADVANTAGES
		• More difficult to replicate • Face-to-face interviews are not anonymous • Telephone surveys are not ideal for complicated closed-ended items or choices • Face-to-face interviews may require payment for participants
Interviews: Focus groups	• Ideal for exploratory research • Better for insights about complex issues and topics • Suitable for studying opinions and attitudes • Group interaction generates new ideas as respondents build on others' comments • Can probe for additional information • Best for small groups (6 to 12 range)	• Not as ideal for collecting sensitive or personal information in some cultures • A few people can dominate the discussions • Responses easily affected by what others say • Minority views often not disclosed • Not as suitable for studying behavior • Time intensive to run • Requires expert skills in leading groups • Small sample sizes in one geographic area • May require payment for participants • Limited to a few topics at a time • More difficult to code responses
Experiments	• Ideal for studying cause and effect explanations • Better control of variables • Easier to replicate • Suitable for collecting quantitative data and doing statistical analyses • Better for achieving internal and external validity	• Ideal for smaller samples but limited generalizability • Experimental laboratory situations are artificial • Narrow range of behavior is measured • Respondents may act in a way because they know they are being studied (demand characteristics of experiments) • Can take much time to run experiments • Equipment costs • May require payment for participants • Ethical concerns about informed consent and harm
Qualitative: Observations and field methods	• Ideal for studying behavior in actual sites • Unanticipated and unexpected findings can be collected • Not limited to structured items on a survey • Allows for respondents' views and perspectives • Behavior and situational factors observed in context and real time	• Limited to smaller samples • Time consuming • More difficult to code observations and responses • Reliability of coding of observations or other content analyses needs to be established • Observer bias can affect what is being observed and how

BOX 1.4 CONTINUED

METHOD	ADVANTAGES	DISADVANTAGES
	• Nonverbal data can be observed and analyzed • Ideal for studying interactions among people • Content analysis can be performed on documents and other written or visual records and artifacts	• Respondents' behavior can be affected by being observed • More difficult to assess opinions and attitudes • Field notes take more time to write and analyze • More difficult to replicate • Ethical concerns about informed consent, role of the participant observer, and potential harm • Not ideal for some quantitative statistical analyses

REVIEW: WHAT DO THESE KEY TERMS MEAN?

Biased questions

Causation versus
  correlation

Content analysis

Describe, explain,
  explore, evaluate,
  and predict

Empiricism

Experiments

Focus groups

Inaccurate
  generalizations

Limited sampling

Needs assessment

Objective and
  intersubjective

Qualitative research

Replicable procedures

Research design

Selective attention,
  perception, and
  retention

Systematic methods

Triangulation

TEST YOURSELF

(answers for this exercise are in the Appendix)

1. What three things must be decided before you can conclude that there is a cause and effect relationship?
2. Using these three factors, what kinds of questions would you ask about autism, or for that matter about storks and birthrates, as described in Box 1.3?
3. What other elements of everyday thinking are evident, and what scientific thinking is needed to look at the relationships described in Box 1.3?

INTERPRET: WHAT DO THESE REAL EXAMPLES TELL US?

1. What are the errors in everyday thinking in the following Letters to the Editor?
   a. "As a parent, I am glad that private and parochial school teachers are not required to go through the training given to public school teachers. If they did, those schools would have the same problems and bad education public school kids face."
   b. "National standardized testing will lead teachers to teach only what is necessary to pass the test. This isn't what teaching is all about."
2. For each of the following academic studies, based just on what is stated here, say whether the main goal of the research is to describe, explain, and/or predict:
   a. Southgate and Roscigno (2009) conducted a study looking at the relationship between elementary and high school students' participation in music programs and their academic achievement in reading and mathematics. Do taking music classes and playing instruments improve students' success in other areas?
   b. Research by LeBlanc and Wight (2000) documents the characteristics of people who serve as informal caregivers of people with AIDS in Los Angeles County and the San Francisco area. Caregivers are more likely to be male, in their late thirties to mid-forties, white, and highly educated but with low income; many are unemployed. However, caregivers related to the patient are more likely to be female, slightly older, a racial/ethnic minority, and earn less money than caregivers who are nonrelatives.
   c. Felmlee and Muraco (2009) sampled 135 adults (average age of 73) to learn about how older men and women might differ in the way they view cultural norms about friendship, especially those related to trust, self-disclosure, commitment, and assistance.
   d. Cyders et al. (2009) measured first-year college students' sensation-seeking personalities and positive moods at the start of the first semester to see if these factors related to their drinking behavior at the end of the second semester. Can understanding certain types of personalities help to prevent later alcohol abuse?

CONSULT: WHAT COULD BE DONE?

One of your friends tells you that she heard from a friend that students living in Mead Hall have the highest grades on campus and suggests that you should move there next semester. When you do so, your grades will go up, she says.

1. How would you respond to this statement?
2. Is this an example of everyday thinking or of scientific thinking? In what ways? What would be the purpose of doing some research on this?
3. How would you respond to the cause and effect statement your friend is making?

DECIDE: WHAT DO YOU DO NEXT?

You are invited to conduct a study on how people develop and maintain friend-ships. The goal is to understand similarities and differences among diverse people. For example, do men and women have the same values about the meaning of friend-ship? How do people of different ethnicities and cultures maintain their friend-ships? Is age an important component of friendship?

1. Give examples from *everyday thinking* that you have heard about friendship forma-tion. What are some errors in making conclusions based on these everyday exam-ples?
2. Provide examples of studies you could do whose purpose is to (a) *explore* friendship formation and maintenance, (b) *describe* the relationships between friendship and people's characteristics (gender, age, etc.), and (c) *explain* the relationships. What kinds of questions would you ask and how would they differ for each type of study?
3. What more would you need to know to declare there is a *cause and effect* relation-ship between friendship formation and specific characteristics (demographics)?

For additional examples, resources, and "test yourself" questions, go to
http://doingsurveyresearch.wordpress.com/

FINDING IDEAS TO RESEARCH

2

Imagination is more important than knowledge. For while knowledge defines all we currently know and understand, imagination points to all we might yet discover and create.

—Albert Einstein, physicist

LEARNING GOALS

Discovering topics to study by searching for research ideas and finding existing studies is one of the goals of this chapter. Learning to write a good literature review is discussed, especially in the context of using theory to guide your research. The chapter also raises the ethical issues involved in doing research. By the end of the chapter, you should be able to search for topics in the library and in computer databases, write a coherent and focused review of the research literature, and note the ethical concerns various kinds of research topics might raise.

You've just been handed an assignment by your boss to gather data for a work-related project about customer satisfaction using a self-administered questionnaire. Or maybe a professor is asking you to develop a research topic for your senior year honors thesis in women's studies. Now what do you do? Where do you even start? You're probably thinking: I wish I had written down all those ideas I've had over the years because now I can't think of anything to study! Figuring out what to research or how to begin an assigned project can be a daunting task for many people. For some, curiosity generates too many broad questions.

Others begin with too narrow an idea that goes nowhere beyond a simple query. What follows are some strategies that can be used to develop a research agenda that is meaningful and focused and that can help in creating a successful research design.

GENERATING TOPICS

Ideas come from many sources, and part of any research design is translating those ideas into reliable and valid ways of measuring them. Developing an idea that is carried out to completion in a scientific research program is a powerful and creative process. To experience taking a topic, constructing research questions, and collecting evidence to support or refute your ideas can be exhilarating. And the first step—generating researchable topics that are unique ideas—is often one of the most innovative parts.

Curiosity and Experience

Where do we get ideas and topics for research? One way is with our own *curiosity.* Look around. Listen attentively to what people are asking, and see what they are doing. Be conscious and questioning of your own *experiences,* and focus carefully on what you read about or hear in everyday encounters. Do these lead you to wonder about something in particular? For example, you learned in a linguistics class that women's body language when listening to others speak appears to be different from men's body language. Are women nodding frequently, providing more verbal signals, and smiling more than men? Or perhaps you heard that voting patterns vary among people from different social classes and ethnicities. Maybe you overhear students complaining about certain of the school's policies. Or you might just simply wonder if studying so much really makes a difference in the grades you get.

One of the first ways, then, of generating a research topic is to use your own curiosity and experiences as a source for further inquiry. But be careful: Personal experiences can result in either a topic so broad that it would be impossible to study ("are men really different from women?") or so narrow that it would be difficult to go beyond a very limited set of questions resulting in the most specific and trivial of findings ("which is the favorite Beatle of my friends' parents?").

Assignments, Theses, and Grants

Of course, another way to find a topic is to be assigned one by a teacher, a supervisor, or some agency offering research grants. Perhaps you get a notice (called a request for proposal or application, or RFP or RFA) stating that *funding* is available for research on alcohol use among the elderly. This sparks your interest, and you begin to develop some questions that you feel need answering in this area.

Similarly, you might be asked to do a *thesis* for completion of a degree in political science, for example. Reflect on the courses you have taken and see if the books you have read or the lectures you remember raise questions that need some systematic analysis. You know that the topic has to be in the area of political issues, so you decide that voting behavior among young adults is something that interests you. Studying voting behavior is quite general and not focused yet, but it is a good start.

Consider a situation where you are an employee at a social service agency that specializes in working with the homeless. The agency is required to file a report at year's end about the services it provided, the kinds of people who have been assisted, and the strengths and weaknesses of the program. Your supervisor *assigns* you to develop a questionnaire that can be used to assess these outcomes. You may not have a choice in the topic, but you do need to make some important decisions about the research design and the kinds of issues to be covered. Again, you start out with a general idea or topic, and you now need to begin focusing more specifically on its multiple dimensions.

Other Research

By reading published academic research, you learn what has been done already and what needs to be accomplished. An *academic article* is a paper that has been reviewed by peers and published in a journal read primarily by researchers and scholars. Normally, these are original reports of research. Such publications as the *Huffington Post* online newspaper, the *New York Times,* or *The Economist* may be sources for original information, but they often summarize and report on research published elsewhere. In such cases, they are secondary sources, not primary ones like academic journals. Your search for further information about your topic should begin with the primary literature of original research.

Sometimes your goal is to *replicate* studies. In such cases, prior research gives you the questions, methods, and information you need to redo the study with a similar or perhaps different sample. An important aspect of science is the ability to repeat results under similar conditions or with different respondents.

Another goal might be to fill a gap in the research by focusing on what hasn't been done. Most published articles conclude with what the researchers couldn't do and with suggestions for additional studies. Sometimes they even suggest particular lines of research that their project has generated and that need further analysis. (See

Box 2.1.) Not only are academic articles a good source for a topic, but they also give you direction once your topic has already been selected. Reviewing the published literature continues even after you design your study; it helps immensely when analyzing data, interpreting the outcomes, and writing up the final report.

Another way of finding research ideas is to make use of *secondary* data sources, such as census data, the Gallup Poll, or the General Social Survey, some of which are available without a fee through the Internet. These sources are called secondary because you are not the primary or first person designing the study or collecting the data. By reviewing the variables and questionnaires from available data sets, you might be able to create your own research project or develop a new way of interpreting earlier findings.

The best data sets have the advantages of probability random sampling; larger comparison groups across various ethnicities, regions of the country, international locations, and other characteristics; and reliable and valid questionnaires that have been tested and professionally developed. Yet, the questions might not be as directly relevant as you would construct for your project, and the data sets might not have enough questions for your particular research goals. Furthermore, the subsamples sometimes aren't large enough for your purposes. But reviewing the methodology and questions used in the collection of publicly available data is a good way of finding research topics and developing hypotheses to study.

BOX 2.1
USING RESEARCH TO GENERATE IDEAS

In an article on the relationship between television viewing and fear of crime, Kort-Butler and Hartshorn (2011) looked at the types of shows people viewed (fictional crime dramas like *Law & Order*, news shows, or nonfictional crime shows like *Primetime* and those on Court TV) and their belief in the death penalty, concern about police being able to prevent crimes, and attitudes toward the criminal justice system. At the end of the article, the authors discuss some of the study's limitations and suggest other avenues of research. For example, they point out that the study was done in Nebraska and that "people in more urbanized states, more racially diverse states, more politically diverse states, and in states with objectively higher crime rates may have different perceptions of crime" (2011: 52). The authors also recommend a longitudinal study and "a comparative content analysis of crime dramas and nonfictional documentary-style crime shows" (2011: 53). Right here are several ideas for a senior thesis, term paper, or funded research project. Look carefully at the end of almost any published paper for creative suggestions for further study.

Serendipity

An idea can also derive from the research we are already doing. It is not unusual to discover a finding that was totally unexpected; this is what is often called *serendipity*. By accident, a result that wasn't anticipated jumps out, and we become intrigued to figure out why this occurred. This leads to a new line of research in an attempt to study this serendipitous finding in more depth. For example, a study is designed to assess the relationship between dropping out of college and grades. Several items of information are gathered during the study, including sex, race, social class, outside work commitments, course load, and major. In the process of reviewing the data, a researcher notices that other items such as whether a student was on academic probation, received low grade warnings, and had meetings with advisors might be interesting to analyze.

Unexpectedly, after further data analysis is completed, it's discovered that some students who received low grade notices tended to drop out rather than get their act together and study harder. Low-income students were more likely to see these warnings as statements verifying that they were not "college material" and could not make it at the university, as they had secretly feared all along. They tended to leave college more than the middle-class students who received the notices and used them as a "kick in the butt" to work harder. Because studying low grade notices wasn't the main focus of the research, it is not possible to go beyond this anomalous finding, but these results can lead to the development of another research project on how probation and low grade notices affect various racial, social class, and gender groups differently.

Whether from astute observations of the world around you, an assignment at work or school, serendipitous findings in another study, or incentives from a funding agency, a topic is generated and the process of refining it to a more manageable project starts. At this point, the primary task is to carve out something that is focused, informative, unique, and fun to do. After all, if you don't enjoy the topic, you are not going to be motivated to do a thorough job!

SEARCHING FOR RESEARCH

A good way of figuring out if an idea or research topic is still too general or too specific is to type some key words into a computer search of resources. You can always use one of the Internet search engines (such as Google), but this will turn up lots of information (some of dubious quality) that is not likely to serve the needs of scholarly research. Imagine, for example, how many millions of responses you'll get if you search for "alcohol and elderly" (I actually got 23.3 million!).

BOX 2.2
LIBRARY DATABASES

There are many resources in the library to guide us in searching for books and articles on specific topics. They are typically available through computer databases, which are lists or collections of information about books and articles organized by such various fields as name of authors, title, year of publication, and key words about the topic. Most of these resources rely on a system using Boolean logic to search for information. Boolean logic is based on dichotomous, or two-category, questions: Is it true or false? Is it an odd-numbered card and a red one? Is it a heart or a diamond? The logical operators "or," "and," "and/or," "less than or equal to," among others, form the basis for the search.

For example, in its simplest form, if we are interested in finding an article about attitudes toward capital punishment among teenagers, then we would search for "capital punishment AND teenagers." Both pieces of information need to be present in an article for it to be selected by the search engine. If we inadvertently typed "capital punishment OR teenagers" then we would get every article written about teenagers and every article written about capital punishment. Research about elderly people's attitudes toward capital punishment would appear, as well as articles about teenagers and movie attendance, for example. Check the "Advanced Search" help feature of these databases for more details.

There are numerous databases available, depending on the subscriptions the library holds. They include *PsycINFO* to search the major psychology journals; *Education Resources Information Clearinghouse (ERIC)* for articles about education topics; *Sociological Abstracts* for sociology; *Social Science Citation Index* for various social sciences; *Political Science and Government Abstracts* and *Public Affairs Information Service (PAIS)* for government and political science areas; *EconLit* for economics; *Anthropology Plus* and *Human Relations Area File (HRAF)* for anthropology; and *MEDLINE* for medical research. LEXIS-NEXIS is a great reference for finding newspaper, magazine, and trade publication articles. More comprehensive databases like JSTOR search hundreds of journals in different fields simultaneously.

Your goal at this point in the research journey is to search for published academic studies on your topic. Using library *databases* and entering the relevant information generates lists of academic articles and studies on your subject. Still, you must narrow your focus. Even when I limited the search of "alcohol and elderly" to published articles using Google Scholar, I was able to narrow the results down to a still unmanageable 707,000. But also be careful not to select too specific a topic such that prior research cannot be found or the outcomes will not be interesting beyond one specific answer (for example, "Did women in Florida under the age of 25 vote in the last election?"). Perhaps uncovering just a few key articles or books on the topic will assist you in narrowing your search.

You also have to begin focusing on what you really want to know; don't ask questions for which you already have the answer. For example, what is it that you actually want to learn about voting behavior and young adults? How frequently they vote, whom they vote for, racial and sex differences in party registration? Once you decide on some other categories of interest, search again using several additional key words, for example, *voting behavior, youth, gender, and political party,* and see what happens now. Ideally, a more reasonable number of articles show up, and you can begin the next phase of reading and analyzing the studies.

Once you have found some well-designed research and academic articles on your topic, a review can provide additional concepts and ideas for further focusing your study. The articles also list other publications to find. Look at the bibliographies and references in these journals and see if there are some articles or books regularly cited. These might indicate a classic study in the field that is important to read, or it may simply provide a new set of articles you hadn't discovered in the database searches.

Searching the academic databases can be followed up with a search of more popular newspapers and magazines, the Internet, and other media to help you understand the popular culture's take on the subject and to suggest other avenues of research. But be careful: A lot of what is in the nonacademic literature may represent inaccurate reporting of results, personal opinion, poor sampling strategies, and the other pitfalls of nonscientific reasoning discussed in Chapter 1. You need to develop a critical sensibility in order to discover the quality material when searching in this era of information overload. Journalism sometimes sounds like social science, but it rarely does more than provide information. Analysis, theoretical interpretation, and systematic evaluation are what make social scientific approaches different from popular culture writing.

LITERATURE REVIEWS

At this point, you should have narrowed your focus and found many articles and books on the subject you are studying. This doesn't mean you have arrived yet at a final topic or set of questions. *Evaluating* the previous research is an important step in developing a high-quality study. Not only does this provide further ideas, but it also generates a set of questions, concepts, and methods relevant to researching your topic.

When reviewing the literature, it is essential to develop a detailed *database,* with index cards, computer programs, or simply sheets of paper in a notebook. Plan ahead and prepare a list of information you need about each publication reviewed. For example, consider the form in Box 2.3. Make copies of the form, complete one for each article or book read, and use it as a way of keeping notes. The complete title of an academic article, its authors, the journal name, the publication date, and the volume and issue numbers are just some of the items required when writing up a bibliography or list of references in a final report, as you can see in the "References" section of this book.

For each academic article or book, attend to the methods used: Who or what is sampled and how many, what questions are asked, what are the variables, how are they measured, what statistics and data analyses are used, are the conclusions linked clearly to the data collected, is there a discussion of the limitations of the study and suggestions for future research? What major findings result from this research? The material in the rest of this book provides information you should use to be a critical evaluator of the methods, sampling, statistics, and overall research design of the studies you read, in addition to teaching you how to do your own survey research.

Once you have assessed the studies, the next task is to decide if you want to replicate any parts of them. If so, you might consider using the same measures (such as questionnaire items with the permission of the authors) or modifying them if you are duplicating the study with a different sample. If you feel something is missing from these studies, pick up on something the authors said they wished they had done, or see if the results suggest a further line of research to take. Then consider which aspects are worth keeping and which should be changed or expanded. For example, if you believe that studying students in an introductory psychology class was a weakness in an otherwise interesting study, then choose a better sampling strategy. If you feel the questionnaire items written for a study on satisfaction in the workplace did not accurately capture what represents satisfaction in the specific workplace you are studying, then modify the questions for your particular sample. Remember, though, that comparisons with other research cannot be exact unless the same measurement tools and methods are used. But this is primarily important only if comparison or replication is your main goal.

A good literature review is necessary to help design the research. There is no need to study something that has been done over and over again, unless you have a unique perspective on the subject. A good literature review is also important in assisting you in becoming more knowledgeable about the research subject. Too often, people jump into a topic and fail to understand the range of issues involved, ignore important cross-cultural or subcultural differences that might impact a study, and make the same methodological mistakes others have. Critically evaluating the historic research record contributes to an expertise that becomes relevant when presenting the findings, contextualizing the research, and responding to queries about the work.

There are several ways of reading the existing academic literature and writing up your review of the key research articles and books. When reading the material that will become part of your literature review, remember that not every article you find is relevant. It is not necessary to write a summary for each item you read, especially if it is not primary research. You should not summarize a study that another author has summarized in her or his research if you haven't reviewed it directly. The author may not be reporting it accurately or may be selectively describing aspects relevant to his or her research and not yours. Try to read original primary sources, not secondary sources.

While reading the literature, focus on the elements you are most interested in evaluating. When you are seeking information about how to measure your variables, for example, compare studies that use different scales and questionnaire items and evaluate the differences. For example, one survey might have a question asking how many years of schooling the respondent has, while another study might ask the education question in broader terms like "elementary school only," "high school graduate," "some college," and so on, as explained in Chapter 4. Other studies might be used to assess the strengths and weaknesses of various sampling strategies (like random sampling versus convenience sampling, as Chapter 5 discusses). Every article or book you read may not always provide information for all the elements of your study, so read selectively and critically.

Use your research questions and goals as a guide to which articles or books are most important in providing you with contextualizing information. Remember, the purpose is not to overwhelm the reader with every piece of research ever done on your topic. Rather, the goal is to provide you with guidelines for your research design and to situate your research in a particular theoretical or research context for the reader.

For writing up a literature review, one good organizing structure is to begin with a brief overview summary of your main research goals, focus, and theoretical perspective. Then, using your research questions, develop a set of categories or themes to discuss the prior literature most related to your goals. Reading the literature involves a type of content analysis in which you seek out thematic links among the articles and books and organize the information into those themes. Reflect on the readings and uncover common threads and differences that run through the work. For example, in a review of the academic literature on friendship formation among elementary school children, you could summarize the studies into categories of age, gender, race/ethnicity, and social class. Or you might notice that the best research focuses on parental involvement in friendship development and the emphasis on value congruence among friends, so you organize your literature review according to those topics. Your goal is to analyze the research that has been done, raise questions about what may be missing from the prior research, and make a case that your research will extend, revise, or replicate what has gone before.

BOX 2.3

CREATING A DATABASE OF REFERENCES

Title:

Author(s):

(For an article) Journal name, volume, issue, page numbers, date:

(For a book) Publisher, city, date:

Summary (sample, methods, variables, questionnaire items, findings):

Direct quotes (with page numbers):

Your opinion (strengths, weaknesses, etc.):

(continue on back as needed)

Some people prefer to write a literature review by summarizing each key article or book, one after another. These kinds of reviews are like annotated bibliographies that describe the goals, methods, and results of each study as related to the overall topic of your research. However, a literature review should have some analysis of the material and not just be a descriptive listing of research studies with brief summaries of the findings. A review should consider specific themes that emerge from the research that has been done and be organized according to the issues, variables, and theories you are using. How generalizable are the results of the past research? Do the findings apply to the sample your study is using? Are certain variables and measures more valid than others for studying your topic? What theoretical perspectives guide previous studies, and how do they relate to your research goals?

When you are summarizing others' research, it is best not to use lengthy quotations directly from the articles or books. Try to paraphrase the information in your own words; use direct excerpts only when necessary to clarify meanings, provide complex ideas, or display the author's unique phrasing and interesting wording. But be careful in summarizing others' ideas and quoting their findings so as not to *plagiarize* information. It is ethically responsible to learn how to properly cite excerpts from someone else's work and not pass them off as your own ideas and words. Look carefully at the research you are reading as a model of how to present quotations and reference books and articles in various disciplines.

When you report someone else's ideas, words, or research findings, you must tell the reader whose they are by using footnotes or in-text referencing. Otherwise, you are giving the impression (intentional or not) that these ideas, words, or research methods are your own creations—in effect, stealing someone else's work. When you directly quote someone else's words, you must also provide the reader with the page number where the phrase appears. In-text citation style is used throughout this book and is discussed further in Chapter 10, about writing up research reports.

The best way to learn how to do a literature review is to notice carefully how the academic articles you are reading do them. Read other research not just to get ideas directly related to your work, but read the literature also as a guide about how to write up your research reports. Consider this example from a study that seeks to understand the relationship between fear of crime and television viewing. Kort-Butler and Hartshorn (2011) open the literature review by stating the main goals of the research and then move into a review of previous research. In a section labeled "Literature Review," they organize other published research into three categories: (1) "Television Exposure and Fear of Crime," which highlights research findings on fearfulness and types of programs watched; (2) "Crime Programming as Info-tainment," which presents articles about this genre of TV crime programs; and (3) "Crime Programming as Ideology," which discusses various theories focusing on research about how "the media, television in particular, is a way through which cultural images about crime are disseminated and reinforced and through which criminal justice policy debates are shaped" (2011: 40). Notice that the authors did not simply summarize a set of academic articles in any random order but instead conceptualized and organized past research studies and theories into three meaningful categories that highlight the central themes and issues related to the key variables of their study on the fear of crime and watching crime-type TV shows.

THEORY AND REASONING

Theories are an important source of research ideas; they typically underlie high-quality research. A *theory* is a set of statements logically linked to explain some phenomenon in the world around us. If a theory is used to generate research ideas about certain behaviors and attitudes, then we are using *deductive reasoning*. Homophily theory, for example, posits that people tend to form social ties with others who are perceived to be similar to them. Based on this, you want to study whether first-year students at your college start to make friends with other students in their residence building who seem similar to them in political and social values. You have deduced a specific research question from a larger theory. On the other hand, if a set of observations or empirical data is used to construct a general system of linked statements, then we are engaging in *inductive reasoning*. In this case, after many observations of students hanging out in the cafeteria and their seating arrangements, you end up with your theory of "birds of a feather flock together." You have used inductive reasoning by going from the particular to the general. Most research involves both processes: A review of the literature tells us what theories and explanations we can use to deduce specific research questions that are then used to get data that form the basis for inducing or modifying a theoretical perspective to explain what was observed and measured. (See Box 2.4 for a classic example of deductive and inductive reasoning.)

A major outcome of investigating a topic through a critical literature review is the discovery of theories and the development of a set of concepts and questions that can be used to test those theories or to create new ones. This set of questions or hypotheses forms the framework of a research design. Learning how to write hypotheses and develop good research questions that can be translated into reliable and valid measures is the focus of Chapter 3.

THE ETHICS OF RESEARCH

When you have selected some specific and manageable topics and ideas to study, the next step is to design a research plan. As you do so, it becomes important to ask

> ## BOX 2.4
> ## DEDUCTIVE AND INDUCTIVE REASONING
>
> Emile Durkheim's *Suicide* (1951 [1897]) is a classic study in which Durkheim *induced* a general theory about social cohesion and how connected people are to communities, based on data collected throughout France in the late nineteenth century. Across different sets of people, those who belonged to more cohesive groups had lower suicide rates, he theorized. Durkheim did not develop a psychological theory about why any one particular individual commits suicide, but rather a sociological one based on specific *aggregated* information (data pooled together from a collection of people). He was interested in explaining *patterns* of behavior across groups and the variables that contribute to an understanding of those patterns. By linking these observations logically into a coherent system of explanation, Durkheim induced a theory about various types of suicide, like egoistic suicide, which is related to a lack of social integration.
>
> As others have done for over 100 years, learning about Durkheim's theory can generate new research questions. Perhaps a study on how people who have large friendship networks tend to be healthier in mind and body, or how a lack of community and shared values contributes to increases in alienation and crime, can be derived from Durkheim's ideas. Going from a more general theory of social cohesion and suicide to a specific set of topics and hypotheses for further study on community cohesion and crime illustrates how research ideas can be *deduced* from theory. Durkheim's contribution to the development of sociology was this creative linking of theory and empirical data.

whether the topic and the act of gathering information are worth the impact they may have on who or what is being studied. You may have generated a wonderful set of questions and topics, but if they put people in difficult and stressful situations, the research should not go forward. In short, before the study begins, it is crucial to reflect on the ethics of doing research about a chosen topic, with the measures you intend to use, under sponsorship (and restrictions) from the agency providing funds, and with the sample of people or institutions you plan to survey.

Every major academic and professional association engaged in research has developed a *code of ethics* to guide the collection of data. Every institution conducting research also sets up an *institutional review board* (IRB) to evaluate the proposed research using those codes before the studies get funded or can start. When human subjects are involved, the guidelines are especially important.

In brief, codes of ethics state that participants should not intentionally be *physically or mentally harmed* and their *right to privacy* must be respected. Potential for harm and threats to privacy arise in several areas and situations, including in the process of sampling, measuring, and analyzing data; disseminating the findings;

and using the data. As described in later chapters, it is unethical when researchers deceptively use inappropriate statistics to distort the findings, distribute portions of the study favorable to their beliefs or the sponsoring agency while concealing unfavorable parts that do not support their ideas, use the results in ways for which they were not designed, and reveal information about specific respondents who were assured confidentiality.

Confidentiality needs to be emphasized when information identifying respondents can be linked to their specific answers and is revealed only to the researchers for the main goals of the project. *Anonymity* can be assured when there is no way of connecting any particular identifying information with the individual person completing the survey. Respondents do not give any names or code numbers linked to their names. Confidentiality is not the same as anonymity: Anonymous information is always confidential since it can never personally be traced back to anyone, but confidential information is never anonymous since it is known who completed the survey.

Although it is not always clear-cut in advance whether a research topic and the questions asked will invade people's privacy or cause mental or physical danger, it is important to discuss the potential impact the study might have on those involved. This is essential in order to eliminate any situations where intentional harm could occur and to inform people what is going to happen in the study so that they may determine how much loss of privacy is at stake and decide whether or not to participate. This is what is called *informed consent.* For example, it would be ethical to tell participants that a questionnaire contains items related to alcohol use and family issues and that they are not obliged to complete items that might disturb them. In this way, should those with a painful family history of alcoholism feel uncomfortable about the project or answering certain questions, they would have the chance to opt out. Of course, this might affect the outcome of the study because it alters the nature of the sample responding, but ethical concerns take precedence.

Similarly, volunteers for research must participate of their own free will. Being part of a captive audience—whether in a classroom or a prison—can be a form of coercion unless there are opportunities to decline involvement. For example, the Code of Ethics for the American Sociological Association (ASA) states, "When undertaking research at their own institutions or organizations with research participants who are students or subordinates, sociologists take special care to protect the prospective subjects from adverse consequences of declining or withdrawing from participation" (for a complete copy of the code, go to www.asanet.org/about/ethics.cfm). However, voluntary participation can affect the outcome of a study if the sample ends up composed only of respondents who are willing to get involved. They may be very different kinds of people from those who declined, and depending on the research topic, this can result in distorted findings.

When there is any danger of physical or mental harm, consent must be given, usually in writing, and it must not be obtained through any form of coercion or misinformation about the project. Researchers must balance the amount of information they need to give with the amount necessary for respondents to arrive at a decision. Sometimes disclosing too much about the research can affect the outcomes of the study. Knowing you are part of the group getting a fake (placebo) vitamin, for example, may affect the results. In no case should the researchers *deceive* the participants about the project. Of course, informed consent may not be needed if the questionnaire does not have the potential for harm, or at least no greater harm than what occurs in everyday social interactions, for example, when asking people anonymously about their favorite books, movies, and other hobbies. Occasionally, *debriefing* people (informing them about the complete objectives and methods of the research) after they participate in the study is a good way of providing information that, if given at the beginning, might have led to biasing the results.

For all research, we need to determine ahead of time whether a project has any potential harm, how we will minimize it if there is any, what mechanisms are in place to guarantee the confidentiality of the data gathered, what benefits the research can have, and how much we will tell the participants before and after the study. This is the kind of information typically presented to an institutional review board that determines whether the design of the study (the sample, measurements, outcomes, uses of the data, consent, and privacy concerns) meets the ethical standards of the profession and sponsoring institution.

Other Ethical Considerations

Some institutional review boards hold that learning how to do research by developing questionnaires for *class projects* may not require human subjects' approval when (a) the participants are informed that the survey is part of a class assignment and list the course and instructor who will see the data, (b) the results are not reported beyond the classroom in any public forum or publication, (c) a statement is included to remind the respondents that their participation is voluntary and that they may skip questions or stop at any time, (d) no sensitive information is collected that can cause mental harm or discomfort in completing the questionnaire, and (e) questionnaires are anonymous. Publicly available data or data that cannot be linked to subjects' identities are also typically exempt from human subject approval by institutional review boards. Because policies vary and change, before you proceed with any research, inquire about the guidelines and code of ethics in effect at your institution or sponsoring agency.

Specific types of research methods may require raising other ethical questions in addition to the standard guidelines presented earlier. For example, with *Internet surveys,* ways of contacting potential respondents should reflect both legal and privacy guidelines that restrict sending unsolicited e-mail ("spam") to participate in a survey. Potential respondents should have a reasonable expectation that they might be contacted for surveys, hold the option of declining participation, not be minors who would normally require parental permission to participate, and be able to have their e-mail addresses easily removed from the mailing lists. Organizations or individuals who send e-mail to recruit respondents should provide legitimate return e-mail addresses and information about the sender that can be verified (see the ethical standards of the Council of American Survey Research Organizations at www .casro.org).

Problems in sampling are a major concern with Internet research. Given uneven distribution of computer access based on age, income, and ethnicity/race, researchers must be ethically aware of the potential for making generalizations about a population based solely on responses to online surveys. Furthermore, the storage of e-mail responses to surveys and the potential for linking answers or direct quotations to someone through e-mail addresses need to be determined in designing Internet surveys.

Another issue facing Internet research is to consider how public the information people provide in chat rooms, discussion boards, and other cyberspace forums really is. Although most of these sites are publicly accessible, people often participate as if their responses were private to other members. Should the participants writing comments on someone's blog be informed that their words are being monitored and analyzed by a researcher? Should the researcher pose as a member of the Internet community being studied or not even announce a presence? How do you get informed consent from people who wish to remain anonymous and who may give false information about their real age, sexual orientation, and ethnicity?

A report on Internet research ethics for the American Association for the Advancement of Science (Frankel and Siang 1999: 9) states, "Guidelines in the physical world allow for deception in the study of human phenomena, providing that the research has considerable prospective scientific, educational, or applied value, that there are no alternative methods for achieving the expected results, that the risks to subjects are minimal, and that sufficient explanation or a debriefing will be given to participants as soon as possible following the conclusion of the research." But it goes on to say that for research in the cyberspace world, "Without a clearer understanding of the benefits and risks associated with Internet research, it may be difficult to justify deceptive practices online." Issues related to informed consent, written agreement to participate, privacy and confidentiality of responses, anonymity, and methods of debriefing for

Internet research introduce new ethical considerations that have yet to be fully developed or understood. Like all matters dealing with the ethics of research, the benefits to the subjects, to society, and to science and knowledge must outweigh any threats to privacy and confidentiality and to the physical or mental harm of the participants and the communities they represent. The debates and issues unique to research online can be explored further at http://eresearch-ethics.org/position.

Quantitative research also introduces ethical considerations uniquely relevant for methods that involve *statistical analysis*. The American Statistical Association's ethical guidelines for statistical practice (www.amstat.org/about/ethicalguidelines.cfm) state that statisticians should do the following:

- Use only statistical methodologies suitable to the data and to obtaining valid results.
- Remain current in dynamically evolving statistical methodology; yesterday's preferred methods may be barely acceptable today and totally obsolete tomorrow.
- Report statistical and substantive assumptions made in the study.
- When reporting analyses of volunteer data or other data not representative of a defined population, include appropriate disclaimers.
- Report the limits of statistical inference of the study and possible sources of error.
- Account for all data considered in a study and explain the sample(s) actually used.
- Write with consideration of the intended audience. (For the general public, convey the scope, relevance, and conclusions of a study without technical distractions. For the professional literature, strive to answer the questions likely to occur to your peers.)

Autonomy, Beneficence, and Justice

The simplest way of summarizing the key principles of ethical research is to invoke a 1979 document created by the U.S. Department of Health, Education, and Welfare known as "The Belmont Report" or "Ethical Principles and Guidelines for the Protection of Human Subjects of Research" (www.hhs.gov/ohrp/policy/belmont.html). This report states the three guiding principles that govern research with human subjects and that should be raised whenever any research is proposed and conducted are autonomy, beneficence, and justice.

Autonomy is the principle of respect for individuals as autonomous agents and protection of those with diminished autonomy (such as the incapacitated, mentally ill, or prisoners). Participants in research must voluntarily participate on the basis of adequate information to consent to their involvement in the project.

Beneficence requires researchers to do no harm, to maximize the benefits to knowledge and society, and minimize the risks and potential injuries to the participants.

Justice refers to fairness in distribution so that no particular group of people is systematically denied equal entitlement to a benefit or selected for participation in a research project because of their easy availability and manipulability, especially when unrelated to the purposes of the study.

This statement from the American Sociological Association's Code of Ethics (www.asanet.org/about/ethics.cfm) says it succinctly and applies to all fields of study. (Just substitute "psychologists" or "anthropologists" or "political scientists" or any other field for "sociologists.")

Sociologists respect the rights, dignity, and worth of all people. They strive to eliminate bias in their professional activities, and they do not tolerate any forms of discrimination based on age; gender; race; ethnicity; national origin; religion; sexual orientation; disability; health conditions; or marital, domestic, or parental status. They are sensitive to cultural, individual, and role differences in serving, teaching, and studying groups of people with distinctive characteristics. In all of their work-related activities, sociologists acknowledge the rights of others to hold values, attitudes, and opinions that differ from their own.

REVIEW: WHAT DO THESE KEY TERMS MEAN?

Academic article
Boolean logic
Code of ethics
Confidential versus
 anonymous
Database

Debriefing
Deductive and inductive
 reasoning
Informed consent
Institutional review
 board

Review of the literature
Search engines
Serendipity
Theory
Voluntary participation

TEST YOURSELF

1. Respondents are given a code number on a survey and reassured that only the researchers will know which code numbers are assigned to specific people and only the researchers will see the responses. Explain whether this is an example of confidentiality or anonymity.
2. Explain the three guiding principles of doing research ethically with human subjects.
3. Now that you have finished this chapter, what does the Einstein quotation at the beginning mean to you?

INTERPRET: WHAT DO THESE REAL EXAMPLES TELL US?

1. An article by Daniel Mears (2001: 3), investigating supposed links between immigration and crime, includes the following in his literature review:

   > Theoretical research on the relationship between immigration and crime has been nominal.... [Clifford Shaw and Henry McKay's 1942] theory suggests that crime is more likely in "socially disorganized" areas marked by high levels of poverty, ethnic heterogeneity, and residential mobility.... In addition to [their] social disorganization theory, at least two other prominent theories have been used to examine the immigration-crime nexus. The first, strain theory, involves a focus on blocked socioeconomic opportunities as contributing to crime and delinquency; the second, cultural deviance theory, derived primarily from [Edwin Sutherland's 1934] pioneering work on acculturation, centers on the idea that certain groups have or develop distinctive cultural traditions that either promote or are accepting of criminal behavior.

   a. How is theory being used to generate ideas for a study here?
   b. Is this deductive or inductive reasoning?
   c. How might these theories relate to the topic of Mears's research?
   d. What kinds of research questions could you generate from these very brief descriptions of various theories?
   e. What are some of the ethical concerns you would raise about such a study?

2. For a study of sexual practices of Asian and Pacific Islander high school students, Schuster et al. (1998: 223) described their procedures for administering the survey:

   > Respondents completed the anonymous self-administered survey during a regular class period and sealed it in an opaque envelope. Survey administrators unaffiliated with the district proctored the classes.... The school district notified parents about the survey and gave them the opportunity to sign a form denying permission for their children to participate. Students could also decline participation, and names of respondents completing the survey were not recorded. Respondents were instructed to skip questions they preferred not to answer.

   a. Using the code of ethics, how do you evaluate these statements?
   b. What other ethical concerns do you feel need to be addressed?
   c. What is the difference between asking parents to deny permission and asking them to give permission? What are the pros and cons of each method?

CONSULT: WHAT COULD BE DONE?

Imagine you have been asked to consult on some research projects. Discuss with the researchers the *ethical* issues involved in the following situations:

1. A questionnaire on alcohol use seeks information about growing up in an alcoholic family.
2. Prisoners in the state penitentiary can get time off if they agree to try a new kind of medication.
3. Students in an intro psychology class get extra credit for volunteering for an experimental study conducted by the professor.
4. Respondents to a survey are asked to tell about their recent sexual experiences with those of the opposite sex and same sex.
5. High school students are invited to complete a questionnaire on drug and alcohol use. Monetary incentives will be offered for completion of the surveys.
6. For a study on nutrition and performance, participants are given high-fat-content meals for several days before being asked to solve some mathematical problems.
7. You are led to believe that you are receiving a new vitamin to help in fighting colds, but later realize you were part of the control group receiving a placebo (a nonvitamin sugar pill).
8. You pose as someone of a different race and gender in an Internet discussion group to collect what people have to say about the meaning of friendship in their lives.

DECIDE: WHAT DO YOU DO NEXT?

For your study on how people develop and maintain friendships, as well as the differences and similarities among diverse people, respond to the following items:

1. Which databases would be most useful for a study of this kind?
2. Develop some key words that you could use in a search of databases to find academic articles on the topic.
3. What Internet sites might provide helpful information on this issue?
4. Make a list of categories and themes you would use to organize the research literature after you have reviewed relevant articles and books.
5. If you are doing such a study, list the databases and the keywords you used, create a bibliography of at least five articles and books related to your specific topic, and write up a summary of the readings organized thematically.

DESIGNING RESEARCH

Concepts, Hypotheses, and Measurement

3

It is a capital mistake to theorize before you have all the evidence. It biases the judgment.

—Sherlock Holmes (by way of Sir Arthur Conan Doyle, writer)

LEARNING GOALS

Central to doing survey research is understanding the idea of operationalization and how to go from ideas to concepts to variables. Learning the various levels of measurement is also essential for analyzing data. This chapter shows how to write hypotheses using independent and dependent variables and how to evaluate the reliability and validity of measures. By the end of the chapter, you should be able to distinguish the different levels of measurement (nominal, ordinal, and interval/ratio), discuss the various kinds of reliability and validity, and create one-directional, two-directional, and null hypotheses.

A fter we select a topic and review the literature, we are ready to begin constructing a research plan. A research design serves as a blueprint for the project and must be detailed when proposing a topic for a thesis or applying for a grant. A *research design* involves several stages: (1) developing concepts that are derived from ideas, theories, or prior research; (2) taking those concepts and translating them into measurable variables (operationalizing concepts); (3) selecting the most appropriate research method to gather data (surveys, experiments, field methods, content analysis, etc.) based on the goals of the project (to describe, explain, predict, explore, or evaluate); (4) choosing a sampling strategy for deciding whom or what we want to study (*the units of analysis*) and over what period of time (longitudinal across time or a one-time cross-sectional study); (5) planning how to collect the data and who will do it; (6) deciding on the relevant statistical and analytical tools to make sense of the findings and observations; and (7) describing plans for interpreting and analyzing the results and writing a final report, article, or policy recommendation. A detailed budget should also be included as part of a research design (especially for a grant or funding agency) and specify everything from the costs of duplicating questionnaires to phone calls, supplies, salaries for researchers and those doing the data collection, computer data entry and software, travel expenses, and other related items. The remaining chapters of the book explore these steps in the research design process as they apply in particular to survey research methods.

VARIABLES AND HYPOTHESES

Creating a research design or plan is essential for carrying out a scientific study. By carefully specifying the steps necessary for researching a topic, we avoid many of the pitfalls of everyday thinking described in Chapter 1. The first important phase in this process is formulating ways to measure ideas and concepts with accuracy and consistency.

Concepts

Research is guided by a set of questions composed of concepts connected to your topic. These concepts may have been uncovered in a review of previous research and theories or developed from your own intellectual curiosity and knowledge. A *concept* is an idea, a general mental formulation summarizing specific occurrences, such as "gender" representing such things as masculinity and femininity, or "age" summarizing specific instances of the idea of time (youth, middle age, elderly). A concept can be defined with a dictionary to produce a common usage of the word, and a search of scientific publications can result in a more suitable definition for the appropriate field of study. Otherwise, the meanings of concepts can be quite subjective and made more difficult to measure, especially for more abstract ones. Some concepts are specific and concrete, such as "height" or "academic major," while others, sometimes called *constructs,* are more complex, abstract, or difficult to define, such as "happiness" or "anomie." Ask many people to define the concept of "love," and you'll get everything from operas to paintings to poetry to warm puppies.

Conceptualization must occur for research to begin, and what we mean by the ideas and terms used in our study should be explicitly stated. For example, if you are interested in studying gender differences in phobias, you need to be clear about your conceptualizations. Psychologists define the concept of "phobia" as a persistent illogical fear, while sociologists define "gender" as the prescribed roles that men and women are to follow in a particular culture. A good way to start a project, then, is to list the concepts you are interested in studying and then find relevant popular and scientific definitions for them.

Variables and Values

The next step is to take the concepts of the research topic and translate them into something measurable. When this is done, these concepts are called *variables* to signify the variation that might exist in the concept. Concepts that have only one fixed meaning, such as the concept of pi (π) in mathematics, are called *constants* since they don't vary. Most concepts have multiple categories or *values* to represent the variability of the concept. "Sex" typically has two values, male and female, and "religion" can vary across dozens of categories. The subjective concept (or construct) of happiness, for example, can be transformed into a variable with values ranging from very unhappy to very happy, and height could range from very short to very tall, or from two feet tall to seven feet tall, depending on how you decide to measure it. Occasionally in a study, what normally is a variable can become a constant, for example, when only women complete the survey. In this case, sex becomes a constant for this particular study and can no longer be used as a variable for further data analysis.

Once concepts have been defined and translated into variables, they require some specification about how they are to be measured, or what is called the process of *operationalization*, as described later in this chapter. Measurable variables form the basis of questionnaire items that guide the collection of data. Questionnaire items are the most specific form of an operationalization and represent what the researchers believe are good indicators of the concept. Chapter 4 focuses on writing questionnaires.

Hypotheses

Central to a research design is the construction of research questions and hypotheses to guide the project. Untested statements that specify a relationship between two or more variables are called *hypotheses*. A hypothesis is a hunch derived from an informed reading of the literature, a theory, or personal observations and experience, and it must be capable of being tested. For example, a study might hypothesize that there is a relationship between employees' job satisfaction scores and the quality of the benefits provided, such as health care, vacation time, and child-care facilities. Each of these concepts is a measurable variable, and a hypothesis spells out the potential association relating them to one another.

Independent and Dependent Variables

One goal of research is to explain why a particular variable varies; why isn't it a constant? Why don't people vote the same way? What is the outcome of a new benefit package in the workplace? Why do people vary in height or hold different attitudes toward equal rights for gays and lesbians? The outcome we are seeking to understand is called the *dependent variable,* and we hypothesize that its variability in our sample of respondents *depends on* particular explanations or causes. The explanations or causes (or predictors, if you are trying to predict the variation in the dependent variable) are called the *independent variables.*

There is nothing inherent in a variable that makes it dependent or independent; it often is a function of the research question or hypothesis we develop. What is a dependent variable in one hypothesis could very well become an independent variable in another hypothesis or study. For example, you might be investigating whether the variation in the number of hours studying (independent variable) in a survey of college students affects differences in grade point averages (dependent variable), but later on you are interested in the reason that study hours differ so much (dependent variable) and whether this variation in your sample is due to involvement in work and extracurricular activities (independent variables).

BOX 3.1
CONCEPTS, HYPOTHESES, AND VARIABLES

Consider the following example from a published academic article. Note how concepts are introduced and defined, how they become variables in hypotheses, how they are operationalized, and which ones are independent and dependent.

A study by Jones and Luo (1999) was designed to assess whether poor people exhibit a culture of poverty mentality and if African Americans differ from whites in their attitudes toward employment, family values, and welfare. They begin their article by discussing anthropologist Oscar Lewis's 1959 thesis of a "culture of poverty" in which poor people cope with their impoverished situations by adapting a set of psychological traits including a strong orientation to the present, early sexual initiation, and a mother-centered family. These represent a set of values and attitudes that are passed along to the next generation through a process of socialization or learning.

BOX 3.1 CONTINUED

After discussing Lewis's ideas and other research about a "culture of poverty," the researchers use these ideas and translate the concepts into more concrete definitions and measures. They (1999: 440) write, "Contemporary culture of poverty explanations identify three distinct mind-sets that differentiate poor individuals from mainstream society: lack of a work ethic, improper family values, and an ethic of dependency."

Note how they take a more abstract concept such as "the culture of poverty" and turn it into a more manageable and specific set of ideas. For their purposes, three concepts are described to indicate a culture of poverty. But the work is just beginning. The researchers next have to define more explicitly what they mean by these concepts.

They define "work ethic" using the concepts and ideas from other research, including whether someone would continue working after he or she earned enough money to support oneself without a job, if that person sees employment as central to getting ahead, and if he or she is aware of how much motivation to work and orientation to the future people have. An "ethic of dependency" is conceptualized as attitudes toward welfare, and the concept of "family values" becomes a shorthand phrase for considering attitudes toward sexual activity outside of marriage, single parenthood, and female-headed households.

With these more specific definitions, Jones and Luo (1999: 444) construct a set of hypotheses, such as "Blacks should not differ from whites in regard to the work ethic dimension" and "Blacks should exhibit more positive attitudes toward mother-only families, out-of-wedlock births, premarital sex, and welfare." They label race as an independent variable and attitudes toward work, welfare, premarital sex, and so on, as the dependent variables. The first research question is stated as a two-directional null hypothesis (no difference between the two categories or values of the variable race) and the second is a one-directional hypothesis, with blacks expected to demonstrate higher attitude scores on the dependent variables. Also notice that these hypotheses are designed for the purposes of describing differences, not for explanation. The outcome would tell us who differs, but not why.

Now comes a key part: The researchers specify the procedures (operationalization) they use to measure these concepts, which are stated as variables in the hypotheses. Jones and Luo (1999: 446) continue, "We use three items to measure work ethic. Our first item ... assesses whether or not an individual would continue working even if she or he no longer needed the income.... Our second item ... assesses whether or not an individual considers employment to be his or her most important activity.... The third item ... [asks] respondents how important it is that a child learns to work hard. This item simultaneously taps the work ethic and socialization aspects of the culture of poverty."

An "ethic of dependency" is operationalized with items measuring whether respondents believe that "individuals should have to work in order to receive welfare" and "welfare benefits should be reduced in order to make employment more desirable." Approval of premarital sex, belief that childbearing should be limited to marriage, and opinion about a single woman's ability to raise children are used to assess "family values." Note that the researchers could have defined "family values" behaviorally—namely, asking respondents whether they had sexual activity before marriage, grew up in a female-headed household, or are single parents. However, they chose to use attitudinal measures to define the concept.

In case you are interested, they found no difference in attitudes toward welfare and family values between whites and blacks, but blacks agreed more than whites that single women can raise a child as well as married couples can. To learn their other findings, look for the article in your library.

Positive, Inverse, One-Directional, Two-Directional, and Null Hypotheses

Hypotheses should be clearly listed before you proceed with a study. However, it is not necessary to have hypotheses before conducting research; sometimes we create simple statements in question form to guide the research. These research questions can focus on describing information, such as "How many respondents in my study are male and how many are female?" or "What is the average score on the reading test for the local school?" Other questions attempt to explain or predict relationships or differences among concepts, such as "Are variations in levels of self-esteem related to height and weight?"

A project on grades and study habits might frame the research goals in any number of ways:

1. I wonder if the number of hours studying per week is related to grades at the end of the semester.
2. If the number of hours studying per week is low, then grades are low.
3. The higher the number of hours studying per week, the higher the grades.
4. The relationship between the number of hours studying per week and grades at the end of the semester differs between men and women.
5. There is no relationship between the number of hours studying per week and grades at the end of the semester.

The first statement is simply a research question providing a hunch about a possible relationship between the two variables. The second statement is an "if ... then" format that specifies a particular direction to a hypothesized relationship. In this case, it is called a *positive one-directional* (or *one-tailed*) hypothesis because, as the independent variable decreases in a particular way (study hours go lower), then the dependent also decreases in the same direction (grades go lower). The statement is considered *positive* because both variables are hypothesized to change together, or co-vary, in the same direction. Conversely, the third statement puts it a different way: Those who study more get higher grades. This is also a positive one-directional hypothesis.

Note that a relationship is not hypothesized about one particular person but about grouped (or aggregated) people. Hypotheses should not be written as "If a person studies less, his or her grades are lower" since the research is not designed to get information about any one individual person. The hypothesis is proposing that across the hundreds of people researched, those who study less tend to have lower grades. Any conclusions or predictions about one person cannot be made.

A *negative or inverse* one-directional hypothesis states that studying for more hours is related to lower grades—a scary thought indeed! It goes in the opposite or inverse direction: As one variable increases across the sample of people in the study, the other variable decreases. If we do not want to specify ahead of time any direction (positive or inverse), then we write *two-directional* (or *two-tailed*) *hypotheses,* such as "there is a relationship between hours studying and grades." Note that the exact relationship remains unstated, and the possibility that it could be a positive one (more hours studying relates to higher grades; fewer hours studying relates to lower grades) or a negative one (more hours studying leads to lower grades; fewer hours studying leads to higher grades) is left open. We have two chances to support or refute the hypothesis.

The fourth hypothesis is an example in which three variables—hours studying, grades, and sex—are involved. The expectation is that the amount of time studying and the grades obtained will not be the same for men and women. It is not stated ahead of time whether the relationship will apply only to men and not to women, or the other way around. Thus, it is a two-tailed hypothesis. Researchers introduce a third variable in a hypothesis when they are interested in multivariate analysis to see if a relationship between two variables holds up under the conditions of a third or control variable, or if they want to assess the impact of two or more independent variables together affecting the outcome variable. Chapter 9 discusses the process of elaborating relationships with multiple variables.

The fifth example illustrates another way of writing a two-directional hypothesis: the *null hypothesis,* which assumes that *no* relationship exists between the variables. It is a common form of stating hypotheses when using inferential statistics, since the logic of research is that you do not directly prove a hypothesis; you either reject or fail to reject (that is, you accept) a null hypothesis. Here's an explanation in terms of logic that may sound like something out of *Alice in Wonderland.* When you accept no relationship, you essentially have failed to prove that the hypothesis is incorrect; you have failed to reject it. You have not demonstrated a statistically significant relationship between the variables. When you reject the null hypothesis, you accept the alternative that there is a relationship between the variables. You reject the idea that there is no relationship. Note that you haven't directly proved a relationship; you have instead disproved that there is no relationship! (See Box 3.2.)

Rejecting the null hypothesis is a statistically significant event and worthy of publication, announcements, and perhaps further funding. It is not necessarily a good or bad thing; no value judgment is implied when rejecting or accepting a hypothesis—you may have found a significant relationship between hours watching television per week and an increase in grades, something you were hoping would not happen—but it is something that gathers attention. The goals of your research guide you in interpreting the results, and the hypotheses assist you in scientifically designing your study.

BOX 3.2
THE LOGIC OF THE NULL HYPOTHESIS

In a sense, a null hypothesis exclaims, "Show me that I am wrong; I say there is no relationship," and you find that either you are wrong ("go ahead and reject my hypothesis; I guess there is a relationship after all") or you have withstood the challenge ("accept the hypothesis of no difference; I told you so!"). But in neither case have you proved the hypothesized relationship directly. Rather, the null hypothesis has failed to be disproved (it is accepted as is, as no relationship), or you have disproved it (it is rejected and a relationship exists; hold a press conference!).

Here's an example in logic: You state, "There is no such thing as a chicken with purple polka dots." You can do a census of all chickens in the world and see if this is so; if there are no purple polka-dotted chickens in the entire population, then you have indeed proved your point. Realistically, you aren't going to check every chicken in the world, so you take instead a sample of chickens (more on sampling in Chapter 5), and lo and behold, none of the chickens in your sample have purple polka dots. Then maybe you find another sample, and once again there are no purple polka-dotted chickens.

You have failed to disprove or reject your statement: You accept it and infer that there are no purple polka-dotted chickens in the world based on your samples. All you can scientifically and logically conclude is that you accepted your null statement that no purple polka-dotted chickens exist. However, you did not prove directly that there exists no colorful poultry like this. Somewhere in some hidden chicken farm, there might be a purple-dotted chicken pecking away. If you found one, you would reject the null, and state the *alternative hypothesis* that there are some. Call in the press for a major news story and T-shirt marketing plan!

Defining and Operationalizing Variables

With a set of questions or hypotheses, list the variables in your study and specify their conceptual and nominal definitions. Be attentive to which ones are independent variables and which are dependent. It is important that the concepts underlying the variables are clearly defined before you go to the next step in figuring out how to measure them. Consider a study on political attitudes and race and the following hypothesis: There is a difference among various racial and ethnic groups about their opinions of the last presidential election or, as stated in the null form, there is no difference in opinions about the election among various racial and ethnic groups. You assume that people's views on the outcome of that contest depend on their race or ethnicity. You list the two variables: racial or ethnic group (the independent variable) and opinions about the election (the dependent variable). This is a two-directional hypothesis since you are not predicting ahead of time which category of the race and ethnicity variable will relate to which set of attitudes about the election.

Now you must define these concepts: What does race or ethnicity represent, and which attitudes do you want to assess? Do you really mean their attitudes, or were you thinking to include a behavioral dimension, such as for whom they voted? Once you decide on the concepts you intend to study and turn them into variables by concretely defining them in measurable ways, the next step is to specify how you are going to measure these variables. The objective is to develop appropriate indicators for the variables and concepts you are studying. A concept can be measured using any number of ways to show its variability. The concept of "sadness" can be indicated by using one "Are you sad?" question in a survey, or it can be assessed through multiple indicators, such as a set of 10 items that form a "sadness" scale. A *scale* is a composite measure combining responses to many questions believed to assess the same concept or variable.

This process of specifying the indicators of a concept is called *operationalization,* because we are describing the operations or procedures it will take to assign values to the variables. Think about the time someone told you that a movie was really interesting, and you asked for clarification on what "really interesting" meant. What you were doing was asking for a definition, asking how your friend measured the quality of the movie, and asking what operations he or she used to arrive at the value of "really interesting" for the attitudes variable. The values for the variable could have ranged from "really bad" to "really excellent," with "really interesting" somewhere in the middle. Or "really interesting" could have been the highest accolade in your friend's measurement scale. The way a concept is operationalized needs to be spelled out in detail in order to be scientific and not lapse into the errors of everyday thinking as described in Chapter 1.

Similarly, if you define the concept of race/ethnicity as a category with which a respondent identifies, then you might operationalize the variable as a person self-disclosing as African American, Asian/Pacific Islander, Latino, White, Native American, mixed race/ethnicity, or other. Or it could be measured as the category that was listed on a birth certificate. The actual item we develop for a questionnaire is composed of the categories of the variable race/ethnicity.

Let's say "attitude toward a political election" is defined as how intensely respondents feel about the election results and whether they perceive them as fair or not fair. Attitude will be operationalized by developing a five-point measure from "strongly agree" to "strongly disagree" to assess what the respondents believe about such statements as "The outcome of the mayoral election was achieved fairly" and "The campaign used too many negative advertisements."

The conceptual definitions and operationalizations are important to describe in a study because if others wish to replicate your research, it is essential that they know the steps you took to measure your variables. When they review the literature and

> ## BOX 3.3
> # CONCEPTUALIZING AND OPERATIONALIZING "A VOTE"
>
> Let's begin with the concept of voting for a candidate. The idea of voting is inherent in our political democratic system, and the concept is fairly concrete. We could then look up a nominal definition in the dictionary, such as "a formal expression of preference." But we still need to spell out how this preference will be expressed. When will we know if someone voted and for whom? How do we operationalize this concept?
>
> The vote counters must specify the operations (the rules or instructions) it takes to arrive at a conclusion that a vote for one or the other of the candidates has clearly occurred. For some, a vote is defined as a chad that is punched all the way through, or by others as hanging by one corner, or just pushed in, resulting in a dimple. Is the intention of a bruised chad good enough, or does a ballot have to demonstrate a complete punch to be measured as a vote? If you color in a circle on your computerized ballot or test but do so sloppily, and the marks go over onto the circle for another candidate or answer, how do people decide what the real vote or selected answer is? In other words, a dictionary definition of the word *vote* has to be translated into a measurable unit with the methods of decision spelled out, or operationalized clearly and with consensus. There must be agreement among different ballot counters on repeated viewings for there to be a reliable conclusion about the intended vote or answer on the test.

find your study, they might agree or disagree with the way you defined and measured the concepts. So long as you are clear with what you did, you have provided readers with enough information to make a fair assessment of your work.

The more abstract the concept, the harder it is to develop an operationalization. How would we measure the constructs of "love" and "alienation," for example? On the other hand, there are only so many ways we can operationalize "income" in a study: Do we ask for income before or after taxes, with or without investments, per month or per year, for an individual or the entire household? But differences in the way we measure concepts can affect the outcomes and interpretation of the findings.

LEVELS OF MEASUREMENT

Operationalizing variables involves specifying levels of measurement and considers how reliable and valid the measures are. For constructing a questionnaire, it is important to understand the many ways variables can be written. Not only can differences in measurement affect the statistics used when analyzing the data (as discussed in Chapter 6), but they also determine the various kinds of information you actually get. Researchers make four general distinctions in measuring variables:

nominal, ordinal, interval, and ratio measurements. In addition, variables are discrete or continuous.

Discrete and Continuous Variables

A *discrete* variable has values that do not contain additional information between those values, such as the number of children in a family. There are only exact numbers of children; there is nothing in between the counting units. Do you know someone with 1.3 children, or 3.75 kids? Type of car owned can be a Honda, a Ford, a Volkswagen, and so on, but there is no category between a Honda and a Ford. A *continuous* variable, on the other hand, has values that have information between those values. Time is measured continuously: The counting units of 1:15 P.M. and 1:16 P.M. contain seconds between them. If your watch has a second hand, you can measure them. In track races, you break down the units between minutes and seconds even further to hundredths of a second, and computers calculate time in nanoseconds. We arbitrarily say, "It's around 1:15" unless you're one of those smart alecks who, when asked the time, responds, "It's 1:16 and 23 seconds" or "We've been dating for 2 months, 3 days, 14 hours, and 6 minutes!"

The distinction between discrete and continuous is useful in interpreting data since it occasionally is reported in the press that, for example, the average family size is 1.86 children. A statistic appropriate for continuous variables (a mean) was calculated for discrete data. A more important distinction, however, is the level of measurement used to operationalize the variable.

Nominal Measures

The simplest, and perhaps least useful statistically, is the nominal level of measurement. *Nominal* variables are discrete measures whose values represent named categories of classification. A measure used to list academic disciplines, such as sociology, psychology, anthropology, history, and economics, is a nominal one. A value or number can be assigned to these categories (called *coding*) as a shortcut summary of the category, but it is not mathematical and no order to the categories is intended. On a questionnaire, for example, the variable "religion" can have the value 1 for "Catholic," 2 for "Protestant," and 3 for "Jewish," or it could be listed as 1 for "Jewish," 2 for "Buddhist," and so on. These are arbitrary numerals that do not represent any type of mathematical quantity. They're like zip codes or telephone numbers. We can't add or multiply them or claim that our phone number is larger and therefore more important than someone else's. If a friend told you the type of car she drives by giving you the category number without telling you the coding scheme, it would make no sense, because it does not represent an actual value or quantity. ("I drive a 3." Huh?)

In fact, we do not need to assign a number to the categories because many computer programs allow for data analysis of text, but it is a convenient way of entering data into a computer program for later analysis. There are fewer keystrokes when typing a "4" than when typing "cultural anthropologist." When there are two categories, the variable is called a dichotomy or a *dichotomous variable.* In some statistical calculations, these two-category variables are called *dummy* variables; for example, the variable "work status" can have one category value "employed" and another value "not employed."

Ordinal Measures

When the category values for a variable are in sequence, the measurement is considered an *ordinal* variable. This is also a discrete measure but one in which the values increase or decrease in a particular order. In this case, it does make a difference which one comes before or after the other. For example, we might ask respondents to indicate their shirt size (the variable) in terms of small, medium, or large (the values). They go in order; we cannot say medium, small, large, and then assign the numbers 1, 2, and 3 to those words since they are not in sequence. If something is in category 1, it must be less than 2, which in turn must be less than 3. Or we could assign numbers the other way around: 1 can be the most, 2 in the middle, and 3 the least—just so long as we keep going in rank order once we start the numbering.

When we are asked to line up in size places, we are using an ordinal measure to determine who is shortest and who is taller and who is tallest. We don't measure with a ruler to get the exact inches; it's just a visual estimate of degree. If we measure a variable with the idea of greater than or lesser than, we are using at least an ordinal measure. For example, you might ask respondents to disclose their yearly income using such discrete categories as "under $25,000," "from $25,000 to $30,000," "from $30,001 to $35,000," and "over $35,000." These categories are in sequence and can be coded with ordered numerals showing increasing degrees of wealth, such as 1 for "under $25,000," 2 for "from $25,000 to $30,000," and so on.

Yet, the ordered number still doesn't tell you the answer without first consulting a coding sheet of information. If someone responded to a question about how much he earned last year by saying, "I earned 3," you couldn't tell what category of income that is just simply by the numeric value. All you know is that he is earning more (or less, depending on the order) than those in the first two categories.

Interval and Ratio Measures

When the value tells you everything you need to know about the variable, you are operationalizing it with an *interval* variable. These are measurements whose numbers are in order with equal size intervals and that have no absolute or fixed zero as a starting point. Temperature, for example, does not begin at zero; in fact zero degrees is not the absence of temperature, it is an actual temperature. Yet the number itself tells you what you need to know. If you ask how cold it is outside, and the response is 45, you do not need to look up what category that numeric value represents. It is an actual number that can be treated mathematically (that is, added and subtracted). It is 10 degrees more (remember it has the properties of ordinal measures) than 35 degrees, and 5 degrees less than 50. However, you cannot say that 40 degrees is twice as warm as 20, or half as warm as 80, since there is no absolute zero point to calculate a ratio.

When there is an absolute zero, it is called a *ratio measure.* Now a zero value represents the absence of something and the start of the numbering system. There are no negative numbers below zero on this type of measure. Thus, zero weight is the absence of all weight; you cannot have a negative weight. Ratio measures allow you to use the other mathematical operations (multiplication and division) to achieve ratios and conclude that someone who is 200 pounds is twice as heavy as someone 100 pounds, or that $10 an hour is half as much as $20 an hour. Note that this is different from a measure that is used to assess the severity of an earthquake. The Richter Scale is not a ratio measure, but one in which the units are based on logarithmic calculations, so that an earthquake of 7 compared to one of 6 is an increase in amplitude by 10. So you can't always assume that a measure using actual numbers is a ratio or interval one.

Ratio measures are so similar to interval ones, except for the presence or absence of zero, that throughout the book, such variables are called *interval/ratio* measures. They are usually continuous measures, although discrete interval/ratio measures also exist, such as number of books in the library, children in households, houses on a block, or any other similar counting situations. The most advanced and important statistics require interval/ratio levels of measurement.

Other Measurement Considerations

Not every measurement falls neatly into these four distinct types. Dichotomous variables technically are discrete nominal measures. However, an order exists when there are only two values: assigning an arbitrary 1 to "home owners" and a 2 to "renters" suggests that renters are "higher" than home owners for the duration of the data analysis in this particular study. Therefore, dichotomies can sometimes be treated statistically as ordinal or interval/ratio measures. Another example is an intensity measure, such as a five-point range (often called a Likert scale) where 1 is "strongly agree" and

5 is "strongly disagree." These are ordinal measures, but most researchers treat intensity scales as interval/ratio measures when the amount of agreement or disagreement is assumed to vary in equal intervals along the points of the measure.

It is ideal to strive for the highest level of measurement when constructing items for a questionnaire. The levels of measurement are themselves ordinal, with nominal the least powerful, ordinal ones in the middle, and interval/ratio the best. Interval/ratio measures have all the properties of the ones below them. We can always recode interval/ratio data into an ordinal measure, but not the other way around. For example, you can ask respondents their ages in terms of years and months (a continuous interval/ratio measure) and then later move them into "young," "middle age," and "elderly" ordinal categories or group them in ordinal age ranges such as "under 20," "from 20 to 30," "from 31 to 40," and "over 40." However, if you operationalize age in either of these last two discrete ordinal ways, you can never get an exact list of ages or calculate a mathematical mean age. If 100 respondents select "over 40" on your questionnaire, you'll never know if all 100 are each 44 years of age, for example, or if they represent a range of ages from 41 to 99.

BOX 3.4
LEVELS OF MEASUREMENT

Consider these excerpts from published academic research and see how researchers decided to operationalize the variables with different levels of measurement:

1. "Relationship status included four groups: participants who reported being single; dating but not committed; in a committed relationship with a same-sex partner; and with a legal status, either a registered domestic partnership, civil union, or a civil marriage with a same-sex partner" (Riggle et al. 2010: 83).
2. "Age and education were continuous variables measured in years. Race was a dummy variable with Nonwhite coded as 1 and white as reference category. Gender was a dummy variable with male coded as 1" (Kort-Butler and Hartshorn 2011: 44).
3. "How much do you like the following types of music: chart music, hip hop, R&B, rock, heavy metal, punk/alternative, dance/house, techno, hard-house, and classic music. Participants rated their responses on a five-point Likert scale with response categories 1 (dislike very much), 2 (dislike), 3 (do not dislike or like), 4 (like), and 5 (like very much)" (ter Bogt et al. 2010: 851).
4. Two items taken from a national survey about health and sexuality: (a) "During the past 12 months, about how regularly did you drink alcoholic beverages? Would you say that it was …" (1 = daily, 2 = several times a week, 3 = several times a month, 4 = once a month or less, and 5 = not at all); and (b) "On a typical day when you drank, about how many drinks did you usually have? Enter number" (Laumann et al. 1994: 660).

BOX 3.4 CONTINUED

For the first example, the variable "relationship status" is a four-category *nominal* measure. The variable relationship status is nominal since there is no order to the categories. Even if the responses were numbered with "dating" coded 1, "single" coded 2, and so on, these are arbitrary numberings that do not have numerical or mathematical properties. "Legal status" could just as easily have come first, with "single" coded 2, and so on. There is no order, so don't let any coding of the categories with numbers fool you! You cannot perform mathematics with these numbers: A 4 assigned to "legal status" does not make it twice that of "dating" coded a 2, even though actual numbers would suggest that 4 is twice that of 2, or two points higher than 2. That is only true with ratio measures. Nominal measures do not have "higher" or "lower" properties.

The second item includes the variables "age" and "education," both measured in years, making it a *continuous ratio* measure. Those who are 40 are twice as old as those who are 20. Respondents with 8 years of education have half as much as those with 16 years of schooling. The numerical values given to the variables are actual numbers that can be added, multiplied, subtracted, and divided to calculate an average (mean) age, for example. The other items about race and gender are called dummy or *dichotomous* variables with two nominal values assigned an arbitrary numeral: 1 is for those respondents who are nonwhite and 1 is also for male, and those who are female or white would be coded 0.

The third example is technically a *discrete ordinal* measure in which the numerical values are assigned in order from "dislike very much" to "like very much" with three other answers in between. It is arbitrary that "dislike very much" is assigned a 1; it could have been a 4. However, once the numbering starts, the remaining categories must be assigned a numeral in order to represent the increase or decrease in opinions. Ordinal values can be described as "more" or "less" and "larger" or "smaller" than those below and above. These comparative properties are central to the idea of ordinal measures. If the categories are viewed as equal-appearing intervals—that is, if the gap from "dislike very much" to "dislike" is seen as the same as that between "like" and "like very much"—then this ordinal measure could be treated as interval/ratio for statistical purposes (as discussed in Chapters 6 through 9). These are often referred to as Likert scales.

The final items indicate two important ways of measuring alcohol consumption. The first one assesses how frequently respondents drank during the past year using a *discrete ordinal* measure. Like the previous example, the answers begin arbitrarily but then follow in order: "Daily" is coded 1 and "not at all" as 5, although "not at all" could have been coded as 1 and "daily" as 5. They must, however, proceed in order once the choice of how to start the categories is made. Note that the intervals between the values are not equal and cannot be viewed as equal-appearing. The gap between "daily" and "several times a week" is not equal to the interval between "several times a week" and "several times a month" and so on. This is an example of an ordinal measure without equal-appearing intervals.

The second indicator of alcohol use asks for the number of drinks consumed on a typical drinking day. The response is measuring quantity with *interval/ratio* values: an actual number that can be treated mathematically. Someone could reply 4.5 drinks; others might say 10 or 1. In any case, this variable assesses quantity whereas the first item represents frequency. Both are needed to indicate alcohol use and abuse, since some people might respond "daily" but then say they have only one drink a day, compared with others who might respond "several times a month" but consume 12 drinks each time. Who then might be the moderate drinker and who might be the binge drinker?

SCALES AND INDEXES

To measure a complex concept, researchers often construct scales and indexes (or indices). These words are often used interchangeably, but some distinctions can be made. An *index* is a set of items that measure some underlying and shared concept. Think about the Dow Jones Industrial Index, which is a set of numerous stocks combined to create a single number. Now apply that idea to a concept such as "happiness" or "prejudice" or "self-esteem." Creating an index is developing a set of items that together serve as indicators of the underlying concept you are trying to measure. Inter-item correlations are mathematically calculated to determine how well the individual items in the set relate to each other and to the overall concept being measured. For example, imagine a "wellness" index created to measure respondents' attitudes toward physical and mental well-being. A set of questions might include such items as "Most days I feel happy about my physical appearance," "I believe exercise is important for my mental health," and "I often feel lonely."

Perhaps a dozen items like these are written, and the respondents are then asked to indicate how well each question represents their feelings on a scale of 1 to 7, where 1 is "not at all" and 7 is "very well." Researchers would calculate an overall total health index score (an interval/ratio measure) by summing together the responses on each item. In this example, with 12 items, the scores could range from a low of 12 to a high of 84. During the index development stage when the questions are pretested, a researcher would calculate how well each item correlates with the others and with the overall score on the health index. Those items that work well together in measuring attitudes toward health are kept, others are dropped, and perhaps newer ones are added. As discussed in the next section, various ways of determining validity and reliability are necessary when constructing items, scales, and indexes.

A *scale* is a set of items that are ordered in some sequence and that have been designed to measure a unidimensional or multidimensional concept. Usually, a pattern is sought from the responses to a set of items, rather than a simple summation of the individual item scores, as with indexes. Take, for example, Guttman scales, in which agreement with a particular item indicates agreement with all the items that come earlier in the ordered set. Respondents would be asked a series of items indicating (1) if they would be comfortable if someone with HIV lived in their neighborhood, then (2) if they would be comfortable with having someone with HIV as a next-door neighbor, and then (3) if they would be comfortable with having a child with HIV in the school where their children attended, and so on. The items are ordered so that agreement with the third one would also indicate a strong likelihood of agreement with the first two statements.

However, people usually refer to a single item that is measured ordinally as a scale, such as questions rated on a "scale" of 1 to 10 or assessed with a Likert "scale." A Likert scale reflects a level of preference or opinion, typically measured on a five-point ordinal scale, such as "strongly agree," "somewhat agree," "neither agree nor disagree," "somewhat disagree," and "strongly disagree." Technically, these individual items are not a scale in the sense of the outcome of the complex process of "scaling." These Likert items or rating scales are often combined to form an index, although such combinations of measures are sometimes called "summated scales." By now, you may be getting the correct impression that the words *index* and *scale* are used interchangeably by many people!

Today, researchers tend to construct mostly indexes (or summated scales) and only occasionally develop cumulative scales like the Guttman one. The word *scale* is more often used than *index,* and the original differences between them have become blurred over time. Whatever it is called, the key point is to remember that the measurement of complex ideas and concepts ideally requires more than a single item. Combining questions into an overall measurement ("scaling") usually requires advance planning, pretesting, and the statistical establishment of reliability and validity. However, it is more common and much easier to create an index by summing a set of items that have been developed to measure a particular concept even after the data have been collected and statistically analyzed. When you learn to write questions in Chapter 4, consider creating a series of items that might work together in measuring a particular concept. The resulting total score is an interval/ratio measure that can be used in many advanced statistics.

ACCURACY AND CONSISTENCY IN MEASUREMENT

We can construct the best measures in the world, but if they aren't accurate and consistent, our findings cannot be trusted. One of the major sources of error in studies is poor quality of the measurements. Operationalizing variables requires attention to two core concepts of research methodology, namely, validity and reliability. Often these cannot be determined until we have already completed data collection, unless we do pilot studies that test our items first before developing a final version or use (with permission) items previously written for other studies that already have demonstrated validity and reliability.

Validity

We all need to be validated, as we say in "everyday speak"; that is, we all need to be recognized and taken for who we really are. So, too, with measurements. They need to be taken for what they really are: Are the operationalizations measuring what they are intended to measure? *Validity* is about *accuracy* and whether the operationalization is correctly indicating what it's supposed to. A ruler would be a valid measure to assess height, but a scale used to weigh yourself would not be a valid measure to assess height. Validity depends in part on what is being studied; the scale becomes an accurate tool when assessing weight in another study. There are several ways of determining if the measures you use are valid: face, content, construct, and criterion validity.

Face and Content Validity. A legitimate, but not very mathematical, way of assessing validity is to see if the measure seems to be getting the desired result. Usually a consensus develops among researchers as to whether a measure is doing what it's supposed to be doing. Take it on *"face* value" and ask, for example, does the questionnaire item about zodiac sign seem like an accurate way of indicating someone's height? It's not likely to be an accurate measure of that variable, so just on the face of it, it's not valid for assessing height. On the other hand, a question such as "how tall are you?" on its face appears to be a valid way of measuring height. Of course, the respondent might not answer truthfully, so face validity is not a perfect way of determining the accuracy of an item.

Content validity is an equally subjective way to understand how well a set of items is measuring the complexity of a concept or variable we are studying. Does the content of the items cover all the dimensions of the idea? For example, does the content of the driving test required to get a license include the range of things necessary to be a safe driver? A driving test that only asked questions about hand signals or did not require a parallel parking demonstration would not be a very accurate measure of driving skills. Its validity would be called into question. Again, consensus among researchers evaluating the measures is used to determine content validity.

Construct Validity. A better way of assessing the accuracy of a measure is to determine its *construct validity*. A construct is an abstract, complex characteristic or idea that typically has numerous ways to measure it. Take, for example, an idea of measuring student satisfaction with the university. This is not as concrete a concept as asking respondents their height.

Imagine we develop several ways of assessing satisfaction. Based on a statistical connection among these various measures (such as finding out that those who are dissatisfied with the quality of the teaching are also dissatisfied with the classrooms, sport and other recreation facilities, and the relationship with the local community surrounding the campus), we conclude that the items developed are measuring the abstract concept of satisfaction with some degree of accuracy. And if we found out that disliking the food in the cafeteria was strong among both those who were satisfied with college and those who were dissatisfied, that item would not be as accurate an indicator of overall college satisfaction and therefore would have lower construct validity.

Construct validity is based on actual results; sometimes it is not achieved until after the data have been collected and statistically analyzed. Then this information is available for the next time someone proposes research on this topic and is operationalizing similar variables.

Criterion Validity. Another good way of determining validity, especially for constructs not easily measured, is to see if the results from an item or set of measures (a scale) are similar to some external standards or criteria. If these other criteria are available at the same time (concurrently) as the new measures are being used, then we establish *concurrent validity*.

For example, results from an item asking respondents their grade point average are compared with the official information stored by the school. Or take a more abstract variable, such as religiousness. We develop a set of questions we feel measure how religious respondents are. We give the items to a group known to be very religious already and statistically assess whether the measure indicates high religiousness for this group. If so, we have established that the items we wrote accurately clarify people's level of religiousness, and we can now use them with more confidence in our own study. The items are measuring what they were developed to measure; that is, they are accurate and valid.

Another type of criterion validity assesses how accurately our measures predict some future, rather than current, outcome. This is usually referred to as *predictive validity*. Imagine we have written a few items for a questionnaire that we feel indicate motivation to succeed in the workplace. We later track respondents to see if indeed they have successfully been promoted, received raises, and been productive employees. We can then statistically determine if the motivation items were accurate predictors of success and, if they are, declare that they demonstrate predictive validity. We or others can now use them with some confidence and accuracy in another study on this topic or to screen potential employees, because we have established their validity.

Reliability

Just because a measure is valid doesn't necessarily mean it is reliable, and validity means little if the measure used is not reliable. Cars are a valid tool to get from one point to another around town; yet, if you've had cars like mine, they have not always been the most reliable. We want to be sure that when we turn the key, the car starts every time. We want some stability. *Reliability* is about *consistency*; it is the expectation that there won't be different findings each time the measures are used, assuming that nothing has changed in what is being measured. For example, we would expect our weight to change little within a few minutes' time, so repeated measures (getting on and off a scale five times in a row) should indeed demonstrate a consistent value. However, if the measurement tool we use to evaluate weight yields inconsistent results, then that cheap scale you bought at the 99-cent store is not very stable or reliable. There are several ways of determining if the measures we use are reliable: test-retest, parallel forms, split-half, and inter-rater reliability.

Test-Retest Reliability. Consider a ruler made of rubber and a 25-inch-long desk. You can use the rubber ruler to measure the length of the desk; it seems to have face validity as a measuring tool because it has inches marked off along the ruler. Because the desk length does not change, repeated measures should yield the same lengths. So you try again and now the ruler tells you that the desk has grown a few inches to 28. You measure one more time and it now is 32 inches long. Either some supernatural event has just occurred or the rubber ruler has been stretched each time it has been used. Results have not been consistent each time you tested and retested with the ruler, and the reliability of the measurement tool is thus determined to be very low. So don't buy rulers made out of materials that stretch!

Similar methods are used to evaluate the reliability of questionnaire items and other types of scales. If we asked respondents to evaluate their fear of crime and a few days later asked them again, only to find different results—and assuming nothing major has occurred in the media or their neighborhoods to alter those feelings in just a few days—we may be dealing with unreliable measures of fear. Many commercial and other standardized scales, such as the Rosenberg Self-Esteem Scale, the SAT, and the Stanford-Binet Intelligence Test, report their test-retest reliability using statistical measures of correlation (discussed in Chapter 7).

Parallel Form and Inter–Item Reliability. We can also determine how consistently a concept or construct is being measured by comparing it to some equivalent measure or set of items, either externally or internally. Many commercial and standardized tests have two versions, and if the same people score approximately the same on both form A and form B, we can say there is some parallel reliability between the alternate versions. This is what you expect when a professor gives a makeup test: you hope it is the same level of difficulty as the original test!

When we compare responses to similar items within a questionnaire to see if there is consistency in the parallel measurements, it is called *inter-item reliability*.

For example, we might ask respondents how many hours they study on average in a typical week and compare this to their answers somewhere else on the questionnaire about the number of hours they study on average in a typical month and other related items. If the number of hours per month is about four times the previous answer, then we can demonstrate some reliability with these measures.

This also illustrates that the more ways we ask something, the more the potential for reliability increases. Imagine using just one question to indicate how happy people are in the workplace, as opposed to asking them to respond to multiple items that are indicators of happiness. Reliability is obtained when you can demonstrate consistency among these multiple measures of happiness.

Split-Half Reliability. A popular way of statistically determining consistency is to look at internal stability by selecting a group of items developed to measure some construct or variable (a scale or index) and then comparing answers within this group. Consider a set of items you wrote to measure the political values of respondents. Let's say there are 10 items. You would split the number of items in half—for example, either the first five and the last five, or the odd-numbered ones and the even-numbered ones, or randomly select two sets of five—and compare the scores of the two halves. If five of the items indicate low political involvement, then the other five should be consistent and also show low political involvement. If so, you can say the scale developed to measure political involvement is a reliable one. A statistic called Cronbach's alpha (α) is often used to assess internal consistency: The closer the correlation coefficient is to 1.0, the more reliable it is.

Interrater Reliability. When researchers use open-ended items on questionnaires or gather information using interviews and other qualitative techniques, it becomes important that the data collected are interpreted in consistent ways. Content analysis and observation field research require some degree of agreement among those who are reading the data or observing. If those coding the data agree, then we can claim there is intercoder or interrater reliability. The interpretations of the qualitative responses are consistent among various coders or readers.

Achieving reliability and validity is part of the process of operationalizing the variables in research questions and hypotheses. And operationalization is ultimately accomplished by the process of writing a questionnaire. The reliability and validity of items and scales in a questionnaire are affected by the skill, creativity, and techniques we employ when designing everything from the survey's format to the wording of the questions. The next chapter takes you through these next steps of the research journey.

BOX 3.5
LOOKING AT RELIABILITY AND VALIDITY

Here are some examples from published research that discuss issues of reliability and validity in the measurements of the variables used in the various studies:

1. Psychologic distress "was assessed through a 5-item measure; items were rated on a 4-point Likert scale ranging from *never* (0) to *many times* (3).... A reliability analysis showed that the scale had strong internal consistency (Cronbach $\alpha = 0.75$)" (Diaz et al. 2001: 928).
2. "Concurrent validation procedures were employed, using a sample of African American precollege students, to determine the extent to which scale scores obtained from the first edition of the Learning and Study Strategies Inventory (LASSI) were appropriate for diagnostic purposes" (Flowers et al. 2011: 1).
3. The Columbia Mental Maturity Scale, published by Harcourt Brace Jovanovich, estimates the general reasoning ability of children from 3.5 to 10 years of age and assists educators in selecting appropriate curriculum materials and learning tasks (retrieved January 2, 2013, from http://psychology.wikia.com/wiki/Columbia_Mental_Maturity_Scale): "Split-half reliability across all levels approaches .90 and test-retest reliability is approximately .85. Concurrent validity with the Stanford Achievement Test, the Otis-Lennon Mental Ability Test, and the Stanford-Binet Intelligence Test ranges from the .30s to the .60s."
4. The 40-item Friendship Quality Questionnaire–Revised "had high internal consistency at both Time 1 ($\alpha = .96$) and Time 2 ($\alpha = .95$) and test-retest reliability of .38 ($p < .01$)" (Kingery et al. 2011: 224).

The first example describes a very typical application of reliability. It combines five items on a questionnaire, each item assessed with an ordinal measure ranging from zero to three, to form a scale indicating psychological distress. The use of the Cronbach statistic (where perfect reliability is 1.0 and no reliability is 0.0; see Chapter 7) indicates that the five-item scale is fairly reliable in terms of internal consistency. *A split-half internal consistency* procedure establishes the reliability of the items by indicating how strongly half of the items correlate with the other half.

The second example focuses on determining whether a published questionnaire used to measure study skills is accurate in measuring actual academic achievement. This is *concurrent* or *criterion validity* since another established measure (in this case scores on the ACT college entrance test) is used as a criterion to assess the accuracy of the LASSI study skill inventory.

The third item presents reliability and validity information about a commercial standardized test available for schools to purchase for developmental assessment purposes. Note that two types of reliability were measured: *split-half* and *test-retest*. Both demonstrate high reliability (correlations approaching 1.0 are considered high). By comparing scores on this standardized test with scores on three other achievement, mental ability, and intelligence tests, researchers were able to demonstrate concurrent validity, although the correlations were moderate.

The last excerpt from a published academic article illustrates how the 40 items that make up this questionnaire on friendship quality showed strong reliability (consistency) among themselves but had a somewhat low reliability when repeated over time.

REVIEW: WHAT DO THESE KEY TERMS MEAN?

Coding
Constants
Constructs
Continuous measures
Dependent variable
Dichotomy
Discrete measures
Hypothesis
Independent variable
Index
Indicators

Interval measures
Levels of measurement
Likert scales
Nominal measures
Null hypothesis
One-tailed, two-tailed
   hypotheses
Operationalization
Ordinal measures
Positive and negative
   (inverse) relationships

Ratio measures
Reliability: test-retest,
   parallel form, split-
   half, interrater
Research design
Scale
Units of analysis
Validity: face, content,
   construct, criterion,
   predictive, concurrent
Variables

TEST YOURSELF

For each of the following hypotheses, say which variable is independent and which is dependent, what the levels of measurement are (nominal, ordinal, or interval/ ratio), and what kind of hypothesis it is (one-directional positive, one-directional inverse, or two-directional null or positive).

1. There is no relationship between education level (1 = high school graduate, 2 = some college, 3 = college graduate, 4 = graduate school) and scores on a scale measuring life satisfaction (scores range from 1 to 10, where 10 = highly satisfied with one's life).

	Which Variable?	Level of Measurement?	Type of Hypothesis?
Independent variable			
Dependent variable			

2. Men are more likely to receive higher hourly wages than women.

	Which Variable?	Level of Measurement?	Type of Hypothesis?
Independent variable			
Dependent variable			

3. There is a relationship between ethnicity/race and political party affiliation.

	Which Variable?	Level of Measurement?	Type of Hypothesis?
Independent variable			
Dependent variable			

INTERPRET: WHAT DO THESE REAL EXAMPLES TELL US?

1. What *levels of measurement* are used for each of the variables described in the following excerpt from a published study on minority stress and mental health?

   Demographic characteristics. Included were age, ethnicity, education, income, and religious affiliation. *Age* is age in years; ethnicity is coded 1 for White men, 0 for non-White men; *education* is a scale of highest grade completed [number of years such as 8 for elementary school or 14 for two years of college, etc.]; *income* is a scale of annual income clusters in varying sized increments ranging from 1 (less than $3,000) to 19 (more than $150,000); *religious* is coded 1 = yes, 0 = not religious.... (Meyer 1995: 44)

2. Here are some *hypotheses* from published studies looking at the relationships between (a) exposure to certain kinds of youth-oriented media and (b) attitudes toward sex and gender roles or (c) drug behaviors. For each hypothesis, label the independent and dependent variables and describe what kind of hypothesis it is: positive or inverse one-directional, two-directional, or null.
   a. "More frequent exposure to youth media—TV, music/music videos, Internet—is correlated with higher endorsement of permissive sex and more stereotypical gender-role attitudes" (ter Bogt et al. 2010).
   b. "There is an association between the rates of use of TV music channels (such as MTV/VH-1) and both smoking and association with smoking peers" (Slater and Hayes 2010).
   c. "Preferences for specific media types—music videos, heavy metal and hip-hop music—are linked to stronger endorsement of permissive sex and stereotypical gender-role attitudes" (ter Bogt et al. 2010).

CONSULT: WHAT COULD BE DONE?

You are asked to serve as an advisor on a project that the local high school is starting. Administrators want to know whether students' self-esteem affects their grades and plans for college. The administrators intend to use some published standardized self-esteem scale and develop new questions about grades and future plans for a survey.

1. What steps would you suggest they take? What should they do first? Begin consulting with them on a research design.
2. What would you advise them to consider when developing items for their questionnaire and when choosing a standardized scale? See if you can find published information for them about some standardized self-esteem scales and evaluate the information about their reliability and validity. (You might look at your library's copy of the *Mental Measurements Yearbook* [*MMY*] or *Tests in Print* [*TIP*], or other databases of standardized tests and scales.)
3. What are some ethical concerns that need to be considered?

DECIDE: WHAT DO YOU DO NEXT?

For your study on how people develop and maintain friendships, as well as the differences and similarities among diverse people, respond to the following items:

1. Write five hypotheses or research questions using 10 different variables. Try to write different kinds of hypotheses, such as a one-directional (one-tailed) negative (inverse) hypothesis, a positive one, or a two-directional hypothesis. Use both null and alternative hypothesis wording.
2. Make a list of all the variables in your five hypotheses, and label each one as independent or dependent.
3. Say how you could operationalize each variable in your hypotheses and what level of measurement you would use.

4 DEVELOPING A QUESTIONNAIRE

> The scientific mind does not so much provide the right answers as ask the right questions.
>
> —*Claude Levi-Strauss, anthropologist*

LEARNING GOALS

In this chapter you will read about the strengths and weaknesses of different types of survey methods. You will also learn how to design a questionnaire: how to write attitude, behavior, and demographic questions and format a survey. Coding responses and preparing data for computer analysis are important skills discussed as well. By the end of the chapter, you should be able to critique poorly written questionnaires, write a good questionnaire for distribution in a small study, and understand the different ways of designing questions and formats for surveys and interviews.

Some years ago, a solicitation letter (i.e., junk mail) arrived in the mail asking for money to "clean up television." I strongly agree that there's much on TV that I would like to see changed, but my list would be to eliminate stupid reality shows and vapid local news programs. This letter had a different focus, though. Attached to the donation card was an "official poll" asking the following questions and expecting only a yes or no answer:

- Are you in favor of cable television now bringing hard-core pornography into your living rooms?
- Are you in favor of television programs which major in gratuitous violence such as murder, rape, beatings, etc.?

- Do you favor the showing of obnoxious and edited R-rated movies on network television?
- Are you in favor of your children being subjected to the presentation of homosexuality as an acceptable lifestyle in prime-time television?

You don't have to believe that the content of contemporary television shows is problem-free to acknowledge that these four questions are slanted in a particularly biased direction. The wording all points to the answer no with little room for disagreement. Who would say yes to *gratuitous* murder (are other kinds of murder then OK?) or favor *obnoxious* movies on TV? You can readily see the problems in orienting the questions in a certain direction through the use of specific words and phrases. This "survey" is a prime example of the kinds of biases that often show up in questionnaires, sometimes even in the most professional kind. The reliability and validity of information gathered with questions like these would be highly suspect and unscientific.

USING QUESTIONNAIRES IN SURVEY RESEARCH

A key element in the achievement of reliable and valid information in survey research is the construction of well-written and manageable questionnaires or interview schedules. Sometimes we are asked to fill out a survey that barely asks enough questions to make clear what we really believe or limits the way we can respond. Other times we get a questionnaire that is too long, and inquiring about too many personal behaviors or private thoughts. No one form or set of guidelines is going to meet everyone's goals, but there are steps that can be taken to minimize the frustrations and noncompliance that can result from biased, wordy, and poorly designed questionnaires.

Before discussing the techniques and ideas that contribute to a reliable and valid questionnaire, let's first consider the pros and cons of using a questionnaire format for collecting data. As mentioned in the previous chapters, a research topic and set of questions or hypotheses, along with the costs and time frame, must determine the choice of methods. Survey research using questionnaires is not ideal for every kind of study. Whether the questionnaires are designed for self-administered surveys, face-to-face interviews, telephone surveys, or computer-assisted forms on the Internet or in an e-mail raises additional considerations in deciding to use survey methods.

Self-Administered Questionnaires

Designing questionnaires for respondents to complete on their own is one of the most common methods of data collection. Questionnaires can be mailed and returned at a later time in person or by mail; distributed to large groups of people in one location at one time, such as in a classroom or at a meeting; or sent through e-mail or placed on a website.

Self-administered questionnaires are best designed for (a) measuring variables with numerous values or response categories that are too much to read to respondents in an interview or on the telephone, (b) investigating attitudes and opinions that are not usually observable, (c) describing characteristics of a large population, and (d) studying behaviors that may be more stigmatizing or difficult for people to tell someone else face-to-face. The anonymity of self-administered questionnaires permits respondents to be more candid, yet researchers do not always know if those responding are who they say they are and if they are answering honestly.

Questionnaires are more efficient tools for surveying large samples of respondents in short periods of time than interviews or other research methods, and with less expense than interviews or telephone surveys. And because they are more suitable to probability sampling (see Chapter 5), generalizing to a larger population is one of the strengths of survey research. However, response rates can vary depending on the distribution method. It is not unusual for researchers to receive only 20 to 30 percent of mailed questionnaires at first. Face-to-face and phone surveys can achieve as high as 80 percent response rates, while e-mailed and online surveys can have response rates as low as 30 percent (Survey Monkey 2011). Low response rates seriously affect how accurately researchers can generalize the results to a larger population. Follow-up postcards or phone calls, reminder slips in mailboxes, e-mail messages, monetary or gift incentives, and other techniques increase the percentage of people who return their questionnaires, sometimes bringing online and mailed surveys over a 50 percent response rate (see also Dillman 2007). Chapter 5 provides more discussion about sample size and techniques for distributing self-administered surveys.

It is less likely that researchers would affect the outcome of a self-administered survey when respondents read the items on their own, compared to a face-to-face interview. This allows for more standardization of the questions and increased reliability, because the researchers are not influencing the responses by clarifying or explaining the items in varying ways to different respondents. However, respondents can answer the questions out of order, jump to the end, or skip around, and thus alter the results by knowing what comes later or by allowing later questions to suggest answers to earlier ones.

Uniform questions and fixed responses (called *closed-ended* items) also limit how much researchers can adjust for cultural differences in respondents, clarify misunderstood items, or explain ambiguously worded questions. How many times have you tried to respond to a question that has response categories that do not reflect how you really behave? Are you being asked to agree or disagree about a controversial topic for which you don't really have an opinion or even know anything about? How valid then are these items in measuring what they are intended to measure?

Minimally, respondents must be able to read the survey, so self-administered questionnaires are not suitable for young children, the visually impaired, anyone with learning and reading disabilities, or people with limited fluency in the language of the survey. In such cases, face-to-face interviews are preferable.

Computer-Assisted and Web-Based Surveys

A popular way of creating and distributing self-administered questionnaires is with computers. Questionnaires (especially short ones) can be sent to respondents by e-mail, or respondents can be directed by an Internet link to a website that hosts the survey, such as the popular Survey Monkey platform. Researchers may create their own website or hire one of many commercial companies to host surveys. Often the programs used to construct the questionnaires are set up to allow for instant coding of the data, thereby eliminating a source of error that often occurs when researchers or their assistants manually enter data from a completed questionnaire.

In addition to knowing how to write a good questionnaire, researchers who do not have the funds to hire outside computer consultants must also become familiar with programming computers or learn software designed to create online surveys. The added time to do this, or the fees charged by commercial companies, may be offset by eliminating the costs of duplicating and mailing surveys or paying people to code and enter data.

However, one major limitation of computer-based surveys is the issue of access. Variations in computer ownership based on race/ethnicity, age, sex, income, and education can dramatically affect the generalizability of findings from computer surveys. Issues of access are a problem if the goal is to make inferences about larger populations. On the other hand, if your goal is to find out who uses a particular website, as marketing researchers do, or if you are interested in assessing satisfaction levels of workers at an organization in which everyone has free access to a computer and the Internet, then using online surveys or e-mail questionnaires is less of a problem.

Tips for designing online surveys are presented later in the chapter.

Interviews

Surveys are often conducted by interviewers who read the questionnaire items to respondents in a face-to-face situation or over the telephone. One version is an *unstructured* or *in-depth interview,* which is ideally suited for exploratory research.

These can be conducted with one person at a time or in *focus groups,* which involve multiple people being interviewed simultaneously in a group discussion. Although there may be a set of queries used to initiate the unstructured interview, interviewers tend to create questions in reaction to respondents' comments in an interactive format. These *open-ended* questions do not allow for standardization of items with fixed responses and can result sometimes in interviewers biasing the data collection. But unstructured interviews are a good way for exploring how people respond to complex topics for which you do not yet have enough specific information. This information can be used to develop a questionnaire for a more structured interview or self-administered study.

This chapter focuses on designing questionnaires that can be used for *structured interviews.* These take place face-to-face or over the telephone using a questionnaire, often called an *interview schedule.* A central issue for interviews is the role of the interviewer, whose style and personal characteristics (such as gender, race, sexual orientation, age) can affect the respondents' answers. Interviews are reactive situations of social interaction in which discussions about personal behavior and opinions with a stranger are influenced by the interview process itself.

Reading standardized items from a questionnaire involves tone of voice, body language, and other styles that may create a different meaning for various respondents. Interviewers are permitted to *probe* (ask for clarification or elaboration when a response is incomplete or ambiguous), especially for questions without a set of fixed responses. They should be trained to follow the wording and write down the responses, or record them in a computer program designed to host the survey, as comprehensively and accurately as they are capable, without adding, modifying, or deleting information.

Structured interviews are in the hands of the researchers who control the flow and ordering of the questions asked (respondents can't skip around on their own), know who is completing the survey, and can employ various visual aids such as charts, cards with lists of responses, and even some self-administered sections for the more controversial questions, which are sealed in separate envelopes. One of the most respected structured interviews, the General Social Survey (GSS) from the National Opinion Research Center (NORC) at the University of Chicago, uses many of these methods.

Response rates tend to be the highest with face-to-face interviews. Not including interviews that occur when stopped on a street corner or in a shopping mall (those must be short and quick), respondents tolerate longer face-to-face interviews than self-administered surveys. However, finding respondents and interviewing them takes more time, results in smaller sample sizes, and costs more than it does to distribute self-administered questionnaires. Agreeing to participate in an interview varies by location (such as cities or rural areas), time of day, number of people working at home or away, educational level, and other characteristics of the household and its members.

Telephone Surveys

Interviewing people by telephone is one of the most popular ways of conducting survey research. It is less costly and time consuming than face-to-face interviews and less subject to the characteristics of those conducting the interview. Many of these interviews are conducted with computers aiding the interviewers and creating a more standardized interview.

Phone surveys have the advantage of face-to-face interviews in probing for information and getting more details through the use of open-ended questions. At the same time, phone surveys have some of the aspects of self-administered questionnaires by creating a more impersonal interaction. On the other hand, respondents might be reluctant to answer questions from a stranger who calls on their personal phone; yet for some questions, they may be comfortable in providing information they would not give when face-to-face with an interviewer whose characteristics (such as age or race) could impact the responses. If they do consent, short phone surveys of around 20 minutes' duration appear to be the maximum many will tolerate. With the increase in use of mobile phones—which in some cases are replacing traditional landline phones—generating representative samples could be a potential problem (see Christian et al. 2010). Sampling issues are discussed in Chapter 5.

Regardless of the method used, questionnaires should be constructed in ways that allow respondents to provide answers candidly, accurately, and consistently. And they have to address the goals and hypotheses of the research clearly and efficiently. Although there are some differences in questionnaire design, depending on the type of survey chosen, what follows in this chapter is applicable to most questionnaires.

CONCEPTUALIZING THE TASK

It often seems like a daunting task to write a questionnaire. Where do you even begin? How long should it be? What kinds of questions do you need to write? How do you go from the concepts and variables in the research questions and hypotheses to a reliable and valid questionnaire? It can be overwhelming, especially if you have never written one before.

One way of beginning is to list the set of research questions and hypotheses proposed and the variables in each of them. Keeping in mind the way in which others may have measured these concepts when you reviewed the literature, begin to extract the key themes, goals, and concepts of the study. Be sure you have at least one questionnaire item for each of the variables in your hypotheses. Operationalizing the concept—that is, making it into a variable and specifying how it can be measured—is the process of developing the questionnaire item. Developing clear and accurate operationalizations for your concepts and variables is the focus of most of this chapter.

Next, make an outline of what you want to cover. Begin by writing down the following categories: behaviors, attitudes/opinions, and demographics. These are the main components of a questionnaire, although your study might not require all of them. Depending on the goals of the research (describe, explain, explore, or predict), you might be interested in learning about the respondents' feelings and opinions about a set of topics (*attitudes*), what they actually do (*behaviors*), or who they are (*demographics*). Each of these main areas should generate a list of specific topics related to the variables in your research questions and hypotheses. Remember, once your questionnaire is written and distributed, it is too late to add new items, so try at this stage to think about as many ways to measure the variables and concepts as you can, even if they are not likely to make the final cut. Sometimes just writing down crazy ideas stimulates thinking about more relevant ones!

Now comes the difficult part: cutting out what you don't need and cannot measure. Is it really important to include zodiac signs in a list of demographics for a study of GPA and study hours? Can you actually measure how many times respondents drank a particular brand of soda two months ago? Do you really need to ask those questions about eye color and favorite food for a study on political views? Are these linked to any previous research or your own hypotheses?

In other words, you need to start winnowing the list down to reasonable items that are clearly relevant to the variables of your topic. As discussed in Chapter 2, a review of the literature provides ideas for important variables to consider, topics that have been underdeveloped, and ways of writing questions similar to the ones you want to ask. Take a look at Miller and Salkind's (2002) *Handbook of Research Design and Social Measurement* for various scales and items used by other researchers. But don't be limited to what has been done before. Writing a good questionnaire is an art, and your task is to be creative and make an engaging survey that people are willing to fill out.

A frequent concern is how long a questionnaire should be. There is no simple answer to this, since length depends on the amount of time you have to write one, how long it takes for a typical respondent to complete a self-administered survey or for an interview to be finished in person or on the phone, how dense or clear you want the format of the form to look, how much money you have to print and distribute the questionnaires, limits imposed by online Web-hosting services, and how many variables and concepts you are trying to measure. The survey should not appear crowded, have a small typeface, or be intimidating. Clear spacing and visually appealing fonts and format also affect the length.

Finally, if you have writer's block and can't seem to get beyond a blank computer screen or piece of paper, consider that what you are about to do when writing the questionnaire is similar to what goes on in any social interaction. The process of developing a questionnaire is, in a way, a conversation between you and the respondents. Before you start to write items, picture yourself in a discussion with a friend about the topic you are studying. Like a conversation, there are often misunderstandings, vague pronouncements, and confusing instructions. Your goal, of course, is to open up the channels of communication and have a clear and focused discussion.

How would you open up a conversation with someone about satisfaction with the programs provided by a local social services agency? You have some research questions that guide your project, so begin with those. Perhaps you would first want to know if a client uses the facilities and how often (behavior). So you might begin by asking this imaginary person in your mind's conversation if he or she is familiar with the social services agency in the neighborhood, and so on. Then you probably want to learn his or her thoughts (attitudes) about the quality of the programs, staff, the resources, and the facilities. And eventually you need to know more about the person (demographics) completing the survey. Use this method to jot down concepts and variables that are relevant to the hypotheses you developed.

MEASURING ATTITUDES AND OPINIONS

Writing items for a questionnaire takes practice; experience eventually helps you learn what works best. Reading other surveys also provides a good source for developing your own. While many of the tips offered in the next several sections apply to most situations, the goals of your research questions and hypotheses ultimately determine the ideal format and wording of your items.

In our everyday lives, we often ask people to tell us their opinions about a movie or the food at some restaurant, how good a particular professor's class is, or what they think the meaning of life is. We are not asking them what they actually do, just what they believe. Sometimes we are interested in evaluating the relationship between people's values and their behavior, or explaining why they acted a particular way using their beliefs about an issue. For example, lawyers may want to predict how jurors will decide a case knowing how these jurors feel about capital punishment, what they think are the reasons for crime, and their opinions about racial issues.

When we want to know what people believe about something, we construct a set of attitude questions. Questionnaires are ideally suited to assess what people report they believe because feelings and opinions are not readily observed and easily measured with other research methods. Keep in mind that what people say they believe is not a substitute for asking them what they actually do. Many times behavior is a much better indicator of what people feel or think about a subject. Respondents might say they are religious and believe in going to services on a regular basis, but when asked to indicate how often they attend, the frequency might be much lower. Which, then, is a better indicator of religious fervor? In either case, you have to ask people directly what they are thinking or doing.

Open-Ended and Closed-Ended Questions

A good way of finding out what people think is to ask them *open-ended* questions. These require respondents to write out or, if doing an interview, to talk about their responses using their own words and ideas. For example, you can succinctly ask on a questionnaire, "Tell me what you think about the issue of gun control." A blank space is provided for their answers. Of course, you will later need to make sense out of these written comments (and the bad handwriting on self-administered written surveys) by doing *content analysis,* a technique that involves distilling key ideas, words, or phrases and coding them according to some system you developed. This raises issues about reliability and whether different people would interpret the content of respondents' answers similarly. Interrater reliability is required to establish some degree of consistency among those interpreting the open-ended responses.

Respondents often don't like to answer too many open-ended items, because it takes more time, despite the advantage of being able to put the issue into their own words. A more efficient method is to develop *closed-ended* items, although it allows for fewer variations in people's responses. These give respondents standardized answers to select from, similar to questions on a multiple-choice test. It's easier and quicker for the respondents to complete; coding responses is simpler for the researchers and more efficient than with open-ended items. However, many feel that closed-ended questions limit the responses people can give, impose on them the researchers' ideas and words, and frustrate those who prefer explaining their answers in more depth.

Filter and Contingency Questions

When we ask people to state their attitudes about a topic, we assume two things: They already have thought about the issue, and they are willing to share their feelings. In many cases, we end up forcing respondents to provide opinions when they do not have any about a specific issue. For this reason, some researchers prefer using *filtering* techniques in which respondents are first asked if they know anything about the issue; if they do, they are then asked their views. This gives a more accurate portrait of those holding opinions about a subject with which they are familiar, and it eliminates those who answer the question without having thought about it before.

For example, pollsters often ask if you know who the candidates are for the presidential race. Then those familiar with the candidates are asked a set of *contingency* questions, which are answered only by those people who responded in a particular way to a previous item, that is, contingent on answering yes to knowing the candidates. They were filtered toward questions applicable only to them, such as how well they know the candidates' positions. Of those who are familiar with their positions, they are now asked how strongly they agree or disagree with the candidates' views. Different outcomes in political polls have resulted when people were asked for whom they plan to vote in the next election compared with results from surveys where filtering occurred by first asking if they are registered to vote and have plans to vote, and then inquiring for whom they will vote.

Intensity Measures

If you are like many people, your feelings cannot simply be characterized in some dichotomous way like black/white, true/false, or yes/no. While we might indeed have a long list of clear preferences and opinions, a good deal of the time we may feel more or less strongly about them. For instance, you might solidly agree that gays and lesbians should not be discriminated against in the workplace, and you might also agree that gun control is not the solution to crime in the streets, but you may not feel as strongly about gun control as you do about equal rights. In other words, it is often not enough to have respondents choose "agree" or "disagree" or "yes" or "no" for a particular set of statements about various topics; you probably also want to know how *strongly* they agree or disagree. What we need is a measure of *intensity*.

Therefore, a good way of writing closed-ended questionnaire items is to measure people's attitudes and opinions with intensity scales. A common one is derived from the scaling technique devised in 1932 by Rensis Likert. It typically makes use of a 1 to 5 rating scale (where 1 = strongly agree, 2 = somewhat agree, 3 = neutral, 4 = somewhat disagree, and 5 = strongly disagree) or sometimes a 4-point scale without "neutral." Many researchers prefer even-numbered scales, thereby preventing respondents' tendency to select the middle neutral ground. However, this forces respondents to choose an opinion when, in fact, they may not really have one. Others like to mimic the original Likert format but with 3-, 7-, or even 10-point scales. *Likert-type scales* could also include response categories of "strongly approve" to "strongly disapprove" or "highly favor" to "highly disfavor" (or any such range of positive to negative opinions).

Regardless of the scale's size, remember to be consistent with the direction of the choices throughout a questionnaire: If 1 is "strongly agree" for one set of items, do not later on make 1 "strongly disagree" for another set of items. If 10 is to represent "high" or "very favorable," then 1 should not be equal to a high ranking for other questions. Keeping the direction the same throughout the survey minimizes errors and confusion.

Whether it is a 3-, 5-, or 10-point scale, respondents are answering not only in terms of agreement or disagreement (sort of a yes/no), but also in terms of how intensely they agree or disagree. By using such Likert-type scales, we minimize the number of times we fall into the trap of writing "do you" questions that simply result in yes/no or disagree/agree responses. They often don't tell us as much as we would like to know. For example, "Do you believe that television engages in gratuitous violence?" or "Do you agree or disagree that grades should be eliminated on our campus?" tend to yield simple either-or answers. To get much better information, inquire about how strongly respondents "agree or disagree with" or "believe in" some statement.

Some researchers prefer intensity scales that are in one direction, rather than the "disagree to agree" range, which allows three directions of positive, negative, and neutral feelings to be reported. Instead, they might ask respondents to evaluate a statement on a 7- or 10-point scale where 1 = low and 7 or 10 = high. The scale goes in one direction without allowing negative or neutral responses. See Figure 4.1 for an example.

Matrix Format

Once you select the topics, writing attitude items and grouping them in a suitable format are the next steps. Rather than asking each item as a question, consider providing statements about which you want people to give their feelings. Instead of asking, "How strongly do you feel about . . .?" over and over again, try a matrix for-

Figure 4.1   One-Directional Intensity Scale

> On a scale from 1 to 7, where 7 is the highest and 1 is the lowest score, evaluate your supervisor in terms of the following:
>
> My supervisor provides many opportunities for me to learn.
>
> 1   2   3   4   5   6   7

Figure 4.2 Matrix Format for One-Directional Intensity Scale

On a scale from 1 to 7, where 7 is the highest and 1 is the lowest score, evaluate your immediate supervisor on each of the following:

My supervisor ...

a. encourages me to do my best at all times.	1 2 3 4 5 6 7
b. is too demanding with my time.	1 2 3 4 5 6 7
c. is easy to work with on projects.	1 2 3 4 5 6 7

Figure 4.3 Matrix Format for Two-Directional Intensity Scale

Indicate how strongly you agree or disagree with the following statements by circling the number that best represents your opinion.	Strongly Agree	Somewhat Agree	Somewhat Disagree	Strongly Disagree
My supervisor ...				
a. allows me to set my own goals.	1	2	3	4
b. needs to develop better communication skills.	1	2	3	4
c. gets along well with his/her employees.	1	2	3	4

mat, which groups several statements together that all require the same responses. Figures 4.2 and 4.3 show two different examples, the first using a one-directional intensity measure, the other using a more typical Likert scale.

Wording Items

Sometimes the answers to an item depend on whose opinions are under scrutiny. If you ask respondents to agree or disagree with "Merit raises should be eliminated for all workers," you will get different results than if you asked, "I feel that merit raises should be eliminated for all workers." The first wording is more about a general belief; the second puts the spotlight on the respondent who might be more reluctant to express his or her own personal view. While there is no simple solution to resolving this, try writing an occasional item in different ways and compare the results. In any case, be consistent throughout the questionnaire: If you start writing items with "I" or "you," don't switch later to more general, impersonal ones, unless you have a good reason to do so. If one goal is to compare your findings to what others have discovered in other surveys, then you should write your items using the same wording as they used.

A common mistake is wording questions about collective groups as if they held a unified opinion. It is difficult for people to answer a question that asks, "My family supports my choice of an occupation" when "family" does not hold one single opinion. Perhaps your mother approves of your goals, but your father doesn't. Does "family" include your brothers and sisters, aunts and uncles, and grandparents? Be specific in writing questions that seek respondents' attitudes about a large group made up of people who may hold various opinions or vary in their behavior.

Try to avoid *negatives* in sentences. It becomes hard for people to know whether agreeing with a negatively worded statement might actually mean they are disagreeing with it. For example, consider "It is not good to stay up late studying before an exam." Disagreeing that "it is not good" means you probably agree with staying up late before an exam, yet trying to figure this out and disagreeing with a negative statement can lead to errors in responses.

Direction of Statements

Mix the *direction* of your statements so that not all the answers for a particular set of opinions lead to "agree" or all lead to "disagree." For example, if you want to learn people's opinions about study habits, word the questions so that they must disagree with some and agree with others: "Getting a good night's rest before an exam is helpful" and "Staying up late studying the night before an exam helps a lot" are both worded without negatives but capture two different viewpoints. A respondent could not really agree to both of these items. This is often better than wording the second one in the same direction, such as "Staying up late studying the night before an exam can hurt your chances for doing well." Both items would be answered "disagree" if the person believes in cramming for exams.

Mixing the direction of the wording is a good technique to eliminate *response bias,* which is a tendency for respondents to answer most questions in the same way, such as simply checking "disagree" for all the questions. It also cuts down on the possibility of respondents providing answers that are *socially desirable,* that is, giving an answer perceived to be culturally acceptable and positive.

Always and Never

Always avoid the use of the word *always* and never use the word *never*! This statement is a good way of remembering the guideline. Unless you wanted to trap a respondent as someone who exaggerates or lies, do not use such loaded words in your questionnaire items. Very few people "always" do something or "never" feel something about a statement. How many people do you know who "never" or "always" feel happy when they enter a crowded room or believe in "never" telling a lie? Even if you are someone who does, there are many circumstances or situations in which your opinions change. It is much better to phrase questions with such words as *most of the time* or *approximately* or *rarely* or *infrequently*. For example, "Approximately how often in the past month have you felt in control of your time? 1 = almost always, 2 = frequently, 3 = rarely, 4 = almost never."

Double-Barreled Items

Avoid writing items that actually measure two things at the same time. These *double-barreled* questions often contain the word *and*. When you ask people how much they like ham and eggs, it may be difficult to answer for someone who likes one and not the other. In the fund-raising survey discussed at the start of this chapter, how do you answer whether you approve of obnoxious *and* edited R-rated films if you are opposed to the editing of movies for television, but don't mind obnoxious content?

Leading and Loaded Questions

It is not uncommon to start inadvertently writing items for a questionnaire in the direction of particular biases. If you ask, "Do you agree that grades should be abolished on our campus?" you are leading respondents in a particular direction and suggesting that they should agree with you. This kind of *leading* question can be eliminated by writing it as a statement and, as mentioned previously, by including it in a set of items that have a range of viewpoints. Or if you want to keep it as a question, ask, "How strongly do you agree or disagree that grades should be abolished on our campus?"

Another version of biasing is to write a *loaded* question, one in which you push in a particular direction by weighing it down with specifically charged or biased words. Invoking authorities or majorities is one such technique to gain agreement: "Most doctors feel that a high-fiber diet is healthy. How strongly do you agree or disagree that we should be eating more fiber on a daily basis?"

People can be led to give different answers by loading the response categories with various phrases. In a 2010 CBS/*New York Times* poll, 70 percent of Americans said they strongly or somewhat support "gay men and lesbians serving in the military," compared to 59 percent who said so when it was worded as "homosexuals serving in the military." When the statement was reworded as "homosexuals serving openly in the military," 44 percent strongly or somewhat supported it, compared to 58 percent when "gay men and lesbians" was used. Not only does "homosexual" result in lower support compared to "gay men and lesbians," but including the word "openly" reduces support even more (see Hechtkopf 2010).

Language

The wording of items should reflect the educational level and reading language abilities of those filling out a survey. Avoid jargon, acronyms, technical terms, and obscure phrases. For example, if you were to ask students, "How effective is the SSC in meeting your needs on campus?" not everyone will know what the SSC is. Write the item using the acronym as well as the full name: "How effective is the Student Senate Council (SSC) in meeting your needs on campus?" Similarly, an opinion question such as "Does the student newspaper obfuscate the truth?" might lead to lots of blank answers or incorrect responses, even though college students should know the meaning of "obfuscate." (Pause here while you look it up in your online dictionary.)

For many respondents, English may not be their first language. How central to your research is attracting diverse respondents? Perhaps translations are needed, or clarification of words that may be too difficult for those with limited English language ability.

Ranking

Sometimes it may be interesting to ask people to rank, or put into an order of preference, a set of items. A list of movie genres could be presented, and the task is to rank order the most favorite to the least favorite by placing the number 1 next to the type a respondent most likes and continuing to the type least liked. Be aware, though, that asking people to rank more than 5 to 10 items can be a difficult task. Furthermore, intensity is not measured, since ordinal measures tell you only which comes before another one, but they do not indicate how far the gap is between the ranks. (See Box 4.1 for an example.)

Knowledge Questions

Many times, the goal of a research project is to uncover how much people know about certain issues, and not just to find out their attitudes. One of the objectives of using filter questions is to discover who is more familiar with particular topics and then to proceed in asking them for more detailed opinions.

However, be careful in trying to get people to answer questions that you yourself could only find out with a larger survey of the population. Avoid "quiz" questions.

BOX 4.1
RANKING VERSUS RATING

Asking people to rank items seems like a good idea. We often go around telling people our favorite smartphone apps, movies, or books. But ranking tends not to tell you how intensely someone feels about a particular topic, only the order. It's the difference between an ordinal measure and an interval/ratio one. The latter always gives you additional information.

For example, if a questionnaire is filled in as follows, you might learn which is the favorite movie genre in your sample, but you don't learn how strongly they feel about each one. Respondents might enjoy comedies but hate all their other choices, even though Mystery and Drama are ranked second and third.

Please *rank* the following movie types beginning with 1 to indicate your favorite:

1	Comedy
3	Drama
6	Action
2	Mystery
4	Horror
5	Western

A better approach is to rate each item on a scale of intensity. Then the respondents would select a number indicating how strongly they like or dislike each of the movie genres. When the ratings are completed, you not only know which one is the most favorite and the least favorite (just compare their ratings to get rankings), but you also know how intensely they liked or disliked it.

For each of the following movie types, indicate how much you like or dislike them by circling the relevant number:

	Like Very Much	Like Somewhat	Dislike Somewhat	Dislike Very Much
Comedy	1	2	3	4
Drama	1	2	3	4
Action	1	2	3	4
Mystery	1	2	3	4
Horror	1	2	3	4
Western	1	2	3	4

Do not expect accurate answers when you test people about what others do or feel, unless it is the goal of your project to uncover what people perceive to be the extent of some particular opinions or behaviors. Usually, the purpose is to see how familiar people are with various issues or to assess their knowledge about some information. So it is not appropriate to ask such things as "How many people at work drink beer weekly?" or "What percentage of the employees made a contribution to charity last year?" unless you want to know respondents' perceptions of how many people drink or what they believe to be the generosity of their colleagues. These "quiz" questions are not substitutes for surveying the attitudes of the population directly.

MEASURING BEHAVIOR

Knowing how people feel about the next major election, for example, or what their opinions are of the latest political candidates is usually not enough. We might also want to know if these opinions translate into action and whether people actually vote what they feel. For many researchers, figuring out how to measure what people do is of major importance when they attempt to make policy, improve working conditions, or evaluate the outcomes of different innovative social service programs. In many cases, finding out what people *do* tells you as much about their values and beliefs as asking them to report their opinions.

We need to remember that measuring behavior with a questionnaire is actually a measurement of what people *say* they do. Unlike research methods that involve observations or other field methods techniques, a questionnaire can only indicate what people remember and what they are willing to tell you about their behavior. Selective memory, selective perception, and a willingness to be candid all play a role in the validity and reliability of the data collected.

Frequency and Quantity

When inquiring about behavior, we must develop questions people are willing to answer with accuracy and consistency. Many of the same suggestions for assessing opinions apply to measuring behavior: Avoid double-barreled questions, minimize the use of "always" and "never," be attentive to language barriers, and eliminate loaded and leading wording.

Your research goals and hypotheses provide direction for what you want to measure. For example, are you interested in the frequency of a behavior or just that it occurred? Do you want to know how many times in the past week your respondents studied, and/or do you want to know how long they studied each of those times? You might erroneously conclude that someone saying he studied seven times this past week, in comparison to someone else saying she studied only twice, is preparing better and is a more serious student. But what if you found out that the person studying seven times a week only read for 10 minutes each of those times while the person studying twice put in two hours at each sitting? Who then prepared better for the exam?

Behavior could be assessed not simply for frequency but also for quantity. Similar to the Likert scales used for attitudes, measuring behavior can employ intensity measures. For example, "Approximately, how often in the past week did you eat in the employees' cafeteria? 1 = frequently, 2 = sometimes, 3 = rarely, 4 = never." You could also get the same information with a more open-ended question and allow for a true interval/ratio measure by providing a blank space instead of a scale with ordinal categories.

Another version might inquire how many times employees ate in the cafeteria this past week according to an ordinal scale of values, where 1 = six or more times, 2 = three to five times, 3 = once or twice, and 4 = never. Note that the options are no longer an interval/ratio measure and you could not calculate a mean number of times because the only data you have are numbers 1 through 4, which do not represent actual number of times. A 4 is not eating 4 times, and 1 could represent 6, 7, 10, or even 14 times in a week. What you have created is an ordinal measure not useful for some of the more advanced statistical analyses that require interval/ratio data for their calculation (see Chapters 6 through 9).

Mutually Exclusive and Exhaustive

It is important to make the values on closed-ended items mutually exclusive and exhaustive. They are *mutually exclusive* when it is not possible for a respondent to select more than one category or value as an answer to the question. They are *exhaustive* when all possible values or categories are provided as responses to the item. All the choices for any given question or item on your questionnaire should offer the full range of possible answers and not overlap, unless the object is to find out, for example, "How many of the following activities have you participated in during the past year?" in which case you would ask the respondents to "select all that apply." But even in such cases, the list of items should be mutually exclusive and not include, for example, "played sports" and also "played basketball."

Or consider the following typical error: "How many days in the past week did you go to the library? 1 = daily, 2 = four to seven, 3 = one or two, 4 = never." Both 1 and 2 could be selected if someone went to the library each day for a week, so the options are not mutually exclusive. "Daily" in this case is the same as "seven." If students went three days, which choice should they circle? Notice that these values are not exhaustive either; they do not provide all possible answers.

Forced-Choice

One type of question that is sometimes used to get people to make a selection is called the *forced-choice* question; respondents are asked to choose between two options, for example, "Do you go to action films or comedies?" Note that the choices are not exhaustive and they are not an indication about how intensely someone likes them or how frequently they see movies of these types. Since a choice is forced, you'll never know if respondents attend another type of movie more frequently or how strongly they like the one they did select. Some respondents may hate both choices, but they were forced to choose the lesser of two evils, not unlike many political elections! You wouldn't know this with a forced-choice format.

Recalling Behavior

Some behavior is not easily remembered, so it is important to ask questions that respondents can actually answer. Do you really remember what you wore two weeks ago Tuesday, or how many times in the past three months you ate French fries? Unless you wore an unusual costume, never ate fries, or indulged every day, it's not likely that you can recall details about ordinary events.

One technique is to give a reasonable time frame in which the typical behavior occurred. Is it likely for respondents to remember how often in the past six months they did something that happens regularly, such as eating pizza or going to the movies? Does it make sense to ask respondents how often in the past week they did something that occurs occasionally, such as going to the dentist? In other words, choose a time frame that makes sense for the behavior you are studying. A good test is whether you yourself can recall the number of times you did that particular activity in the time frame you decide to use. Pretesting questionnaire items is also an important step for uncovering problematic time frames.

Response Bias

As with attitudes, some respondents are prone to exaggerating the truth or giving socially desirable answers about what they do or how they acted in particular situations. To assess whether there is a response bias, researchers often build in "trap" questions that are likely to pick up those who tend to exaggerate or lie. Purposely using words like *never* and *always* is one way to catch social desirability. Another is to include unlikely choices among the closed-ended items. For example, if you are asking respondents how many of the following books they have read, be sure to include the name of a book that doesn't exist:

During the past year, which of these books have you read? (Circle all that apply.)

1. *A Clockwork Orange*

2. *Black Beauty*

3. *The Scarlet Sweater*

4. *The Color Purple*

When a questionnaire shows evidence of response bias, you need to decide whether to eliminate the exaggerated items or the entire survey from further analysis.

DEMOGRAPHICS

Critical to understanding human behavior is knowing how opinions and behavior vary across different categories of people. Demographic items are those that provide information about the respondents completing the survey. Typical are questions about gender, race and ethnicity, age, income, education, and occupation. Other relevant questions can include political affiliation, religious upbringing, sexual orientation, and city or country of residence.

Many people have the tendency to ask for personal demographic information at the start of a questionnaire. But professionals recommend that these questions and any other easily answered ones come at the end of a self-administered questionnaire. After some fatigue or impatience in having to fill out a survey, most respondents prefer to end the session just checking boxes that are simple to answer. In the meantime, you've gotten answers to many of the important behavioral and attitudinal items that came first on the survey.

Relevant Questions

Any list of possible demographic characteristics could get fairly long, so it is important to limit items to the ones most relevant for the project and for investigating the specific variables in your research questions. If you are researching the number of hours students study and their grades, are political views really important to know? Do Democrats and Republicans study at different rates? While that might be a fun and unusual thing to learn, is it central to your project, or have others found such a relationship in the past that needs replication? If not, then focus on the most important items as a way of maintaining a questionnaire of reasonable length.

When developing a list of relevant questionnaire items, be clear about what aspects of the characteristic you need to know. For example, asking respondents

BOX 4.2
EXAMPLES OF ITEMS FROM THE GENERAL SOCIAL SURVEY (GSS) INTERVIEW

The National Opinion Research Center at the University of Chicago conducts one of the most important ongoing studies of social behavior. Here are two examples from their interview schedule:

1. Are you currently married, widowed, divorced, separated, or have you never been married?

Married	(ASK A, B)	1
Widowed	(ASK A)	2
Divorced	(ASK B)	3
Separated	(ASK B)	4
Never married	(GO TO Q.18)	5

ASK ONLY IF CURRENTLY MARRIED OR WIDOWED:

A. Have you ever been divorced or legally separated?

Yes	1
No	2

IF CURRENTLY WIDOWED, SKIP TO Q.18; IF CURRENTLY MARRIED, SEPARATED OR DIVORCED, ASK:

B. Have you ever been widowed?

Yes	1
No	2

2. How often do you attend religious services?

Never	00
Less than once a year	01
About once or twice a year	02
Several times a year	03
About once a month	04
Two to three times a month	05
Nearly every week	06
Every week	07
Several times a week	08

what their religion is does not indicate how religious they actually are. Do you want to know their religious opinions (such as belief in a higher being), their religious behaviors (how often they attend religious services), the religion they were raised in, or their current religion? These are four different questions that result in four different ways of characterizing respondents in your sample. Similarly, questions about political views (conservative, moderate, or liberal), political behavior (voting, signing petitions, etc.), and political party affiliation (Democrat, Republican, Independent, Green, etc.) do not give you the same information.

The range of choices for a particular question should be appropriate for the sample. For a survey given to high school students, a measure of age that has the following values would no longer be a variable because almost everyone would circle the first option: 1 = under 18, 2 = 18 to 25, 3 = 26 to 33, 4 = 34 and over. Or a study of retirement could also result in a constant, because almost everyone would select the last choice. In these cases, the goal of describing or explaining differences in attitudes and behaviors by age could not be achieved.

Choosing Levels of Measurement

As with behavioral and attitudinal items, you can decide to word some questions as interval/ratio measures or as ordinal ones. For example, asking respondents' ages can be done in several ways. An open-ended question would result in an actual number that can be used to calculate a mean: How old are you? _____.
During the data analysis phase, you could create an ordinal measure from this and collapse the ages into category ranges. This is called *recoding* when you instruct the computer program to take, for example, the numbers under 21 and assign them to category 1, take the numbers 21 to 25 and recode them 2, and so on. But ordinal data cannot go the other way and be made into interval/ratio data.

You could also ask people to provide their birthdate and have the computer calculate the exact age to the day, resulting in a very specific interval/ratio measure. On the other hand, an ordinal version would have category ranges like this: 1 = under 21, 2 = 21 to 25, 3 = 26 to 30, 4 = 31 to 35, 5 = 36 and above. Notice in this last version that the categories are mutually exclusive and exhaustive. When providing categories of numbers, such as age, weight, income, or height, each response should contain ideally the same range. This is better than writing the item in the following way: 1 = under 21, 2 = 21 to 25, 3 = 26 to 28, 4 = 29 to 33, 5 = 34 to 52, 6 = above 52. When the categories are of equal size, you can treat them as equal-appearing intervals and thus use some of the more advanced statistics that are normally restricted to interval/ratio measures.

Sensitive Items

People feel more comfortable answering certain sensitive questions if they are written in categorical ranges rather than as specific interval/ratio numbers. Some might not want to tell you their exact ages but would prefer saying "between 30 and 40" instead of 39; others might be more candid about how much money they earned last year if they could choose "between $45,000 and $55,000" rather than filling in a blank line with "$53,250." The more sensitive the demographic questions (and this depends on who is in the sample), the better it might be to measure these items using ordinal categories rather than trying to achieve specific interval/ratio numbers, even if these allow for more sophisticated statistical analyses. The ultimate goal is to minimize the number of people leaving answers blank, and the trade-off needs to be evaluated before constructing the final wording of the questions.

Mutually Exclusive and Exhaustive

Demographic questions also need to be mutually exclusive and exhaustive. Be sure to include the category "other" in questions for which you cannot anticipate all possible answers. It is not practical to list every religion, for example, but it is important to list those that you think will be most represented in your sample. Buddhism, Islam, or Hinduism might be relevant for some locations, but elsewhere "other" might be sufficient. If you get very large numbers of people choosing "other," this tells you that you may have left out a major category. Including a blank line after "other" helps you see common answers that can then be coded. Be sure the choices are mutually exclusive; for example, including "Protestant" along with "Christian" in the list of religions can result in both categories being checked.

FORMATTING THE QUESTIONNAIRE

Creating a readable, visually pleasing, and comprehensive questionnaire requires practice; this means making multiple drafts, pretesting items, conducting pilot studies, and fine-tuning the final format. And it requires making decisions and evaluating the trade-offs, such as balancing the desire to get lots of information with the need to develop a questionnaire that does not take a very long time to complete and does not seem visually crowded.

Contingency and Branching Questions

The format of a questionnaire and the use of arrows, boxes, and other visual tricks to filter people in skipping questions or branching off to other sections help the respondents flow through a survey smoothly. In general, too many contingency items in a questionnaire can create confusion and missed responses, but surveys need to plan for all possible outcomes and types of respondents.

Be sure to test out the branching or skip logic by taking the survey yourself as if you were multiple people, each with a different response. For example, the survey asks how often respondents drank an alcoholic beverage during the past month. If some answer "none," then they should not drop down to the next question asking which ones and how many. They need to be redirected around the next set of questions and told where to pick up the next relevant items. Read over these questions first as someone who answers "none" and then again as someone who answers one or more times. This way you can see if the branching is correct.

Without branching, people automatically go to the next numbered item, so it often is not necessary to say, for example, "If yes, skip to the next question." They will go there anyway. It is more important to tell respondents who answered no where they should continue the questionnaire. For interviews, it may, however, be necessary to remind the interviewers what to do when respondents say yes or no, even if the next step is to go to the next question. Branching directions in interviews are for the sake of the interviewer, not the respondent, as illustrated in Box 4.2.

With Web-based and other computer-assisted questionnaires, branching is smoother, and errors in completing irrelevant items are virtually eliminated. Those filling out the forms could not erroneously answer a section inappropriate for them because it would never appear on the computer screen. But in written questionnaires, branching can be done in words, such as "if none, skip to Question 3" (see Figure 4.4).

You can also do this visually with arrows and boxes and channel the flow where items in a box are contingent on particular responses to a question (see Figure 4.5). Those answering "none" to item 1 would go to the next question automatically, but those selecting the other choices branch out to a box applicable only to them.

Ordering of Items

Begin a questionnaire with the most interesting topics in order to pull readers into the survey and provide an incentive to complete it. In interviews, it often is better to begin with simpler, more routine questions, including demographics, in order to break the ice and establish a positive interaction and conversation with the respondents. It would be a bit abrupt to begin an interview with the most sensitive and controversial questions. As you start to organize your items and group them accordingly, keep in mind that earlier questions can affect how people interpret or respond to later questions. The order of the response categories can also influence

Figure 4.4   Examples of Branching Questions

1. During the past month, how often did you drink alcoholic beverages?

None 0   (if none, skip to Question 3)

Once or twice 1

Three to five times 2

Six times or more 3

2. For each of the following beverages, please indicate approximately how many you drank during the past month.

       a. Beer _____

       b. Wine_____

3. How many days during the past week did you study for more than one hour?

None 1

One to three days...............................2

Four to six days 3

Daily 4

Figure 4.5   Example of Branching Question Format

1. During the past month, how often did you drink alcoholic beverages?
    none 0
    once or twice 1
    three to five times . . 2
    six or more 3

1a. For each of the following beverages, please indicate approximately how many you drank during the past month.
A) beer_____
B) wine _____

2. How many days during the past week did you study for more than one hour?
    none 1
    one to three days. . . 2
    four to six days. 3
    daily. 4

answers. A study found that when respondents were asked whether the country had a "health care problem" or a "health care crisis," 55 percent selected "crisis." However, when another comparable sample was surveyed and the order was changed to whether the country had a "health care crisis" or a "health care problem," 61 percent said "problem" (Budiansky 1995). The results demonstrate that order makes a difference. Similarly, if you have a set of questions asking people, for example, about their views on a variety of problems in their community (such as unemployment, street cleaning, taxation, and related items), and then follow this section with a question about how well their political leaders are doing and their confidence in them, the results may be biased by having presented all the problems first. If you initially inquire about their attitudes toward politicians, and then seek their opinions about various problems, there could be a different outcome. It is important then to be aware of the order of your questions, which sections come before others, and how these might affect the responses. With time and money, some researchers pretest the order of items to see if there is a difference or develop reliable parallel forms using different ordering of questions and compare the results during the data analysis phase.

Sections and Numbering

Dividing your questionnaire into sections and numbering all of your items are important formatting considerations. People will follow the numbers and know where to go next and will overlook any items that are not clearly numbered. It helps if the survey has numbered or lettered sections, each focusing on a particular set of questions and each with its own brief instructions or description of what is being asked.

For example, Section A might include the attitude questions, Section B the behavior items, and Section C the demographics, if all of these are relevant for the project. Items could be numbered consecutively within each section or continue across the various parts of the survey. This latter style is easier to code and eliminates any confusion about which item number 4 you mean when you direct respondents in a contingency or branching question—the one in section B or C?

Questionnaires are interactions between the researcher and the respondent that should reflect the kind of conversational style you would use when talking with someone. Just as you try to avoid wandering from one topic to another, avoid jumping around in a questionnaire. The goal is to create a smooth flow between sections that does not confuse the respondents and result in incomplete or improperly filled out surveys.

Try to place items together on a questionnaire in sections that are about the same things or require similar kinds of responses. Respondents find it difficult to focus on a particular issue, then jump to another, only to return to the original topic. However, there may be times when it is necessary to mix together various issues and formats, in order to avoid having the respondents figure out the goals of the survey, or to alleviate repetition and boredom, thereby resulting in biased answers.

Be sure not to have page breaks in the middle of an item with the question at the bottom of a page and the response categories on the top of the next page or some responses on one page and the rest on the next.

In addition, each line should have only one question. It is easy to overlook items, especially if they are not numbered and they look like this:

> How old are you? _____   What is your gender? Male 1   Female 2

Instructions

Begin a survey and each section with a brief and clear set of instructions. Avoid giving away everything about your study, since telling too much might affect the responses and the outcome of the research. However, ethically you need to disclose enough information for the respondents to arrive at an informed decision about whether to complete the survey or not.

For example, if you disclose in the instructions that you are studying the relationship between job satisfaction and salary, this could lead some people to answer with that in mind—some might think, "Because I have low income, I will say how dissatisfied I am." An alternative set of instructions could simply state, "We are interested in your candid responses about how you feel about your job."

Don't say, "Please be as honest as possible," since that implies respondents are dishonest people; the word *candid* invites people to be open without the baggage of the word *honest.* Be attentive to other loaded words and phrases in the instructions.

It is ethical to inform respondents whether their answers are *anonymous* (no names or identification numbers are given that might be linked to individual respondents) or *confidential* (names or code numbers are given, but the responses will not be revealed about or connected to any one particular respondent).

If the surveys are not being completed online or in a group setting during a scheduled time and place, be sure to give respondents information about the *due date,* typically placed at the end of the survey, in a cover letter, or in opening instructions. Do not make it a long time away, or the surveys might sit around for weeks and never get completed. And remind them how and where the surveys should be *returned,* such as sent back in the postage-paid envelope, dropped off at a box in front of the personnel office, or deposited in a mail slot. Each of these factors can seriously affect response rate.

Try to be consistent in what you instruct people to do with items. If you have them circle their answers in one section, do not ask them to use check marks in another similar section, or to circle the number next to their answer in a later section. Fewer errors will occur if respondents are not confronted with numerous styles and response formats. However, used sparingly, different response styles can help relieve some boredom and inattentiveness when completing surveys.

Some researchers like to have respondents use brackets (with word-processing programs actual boxes (□) or other shapes (e.g., ◇) can be created) in which an X or check mark is placed, whereas others prefer to have respondents circle the precoded number. Whichever the case, be clear in your instructions and consistent in the format.

Figure 4.6 shows several different styles that can be used, but note that the third example, the one with blank lines, is more susceptible to ambiguity and should normally be avoided. A check mark on one line hurriedly done might overlap the nearby lines and responses, resulting in confusion about which answer was selected. And the second example, with the brackets to be checked, is clear and easy to complete but is not numbered for easy coding. The precoded format in the first example is preferred by many researchers.

Figure 4.6   Formatting the Questionnaire: Three Styles

A. For each of the following questions, please circle your response.

1. I believe the governor is doing a good job running the state.

Strongly agree 1

Agree 2

Disagree 3

Strongly disagree 4

2. For each of the following questions, please indicate your response by placing an X in the appropriate box.

1. What is your current relationship status?

[   ] Single

[   ] Married

[   ] Separated/Divorced

[   ] Living with someone of the opposite sex

[   ] Living with someone of the same sex

3. For each of the following items, place a check mark next to those activities you did in the past two weeks.

_____ went to the movies

_____ surfed the Internet

_____ studied research methodology

ONLINE SURVEY DESIGN

Most of the ideas, tips, and examples provided throughout this chapter are also relevant for surveys designed to be taken on a computer, although you may be limited by the computer program in terms of how you format or what designs you can use for the questionnaire. Computer-assisted questionnaires are popular, especially because the tasks of having people code the data and enter the results into a program for data analysis are minimized. Potential errors are eliminated, and often instantaneous results are available. However, there are some added costs in purchasing and learning a commercial program to design the survey or in contracting with one of many available Internet companies that host and design online questionnaires for a fee. In addition, if you plan on developing your own survey from scratch, hiring computer specialists to write the code, compile the data, and produce statistical results requires funding. Several online survey-hosting sites are free but limit the number of questions and respondents you can have. See http://lap.umd.edu/survey_design as one example, for a guide to writing online surveys.

Central to developing online surveys is learning to create smooth and clear navigation through the pages of the questionnaire. Most of the branching or contingency questions automatically take respondents to another section without much notice or confusion. Skipping items should be done by the computer program and not by the respondent, as is typical with self-administered paper surveys. Writing questions and designing the survey to capture branching and skip patterns will not require the instructions described earlier in the chapter, such as, "If yes, skip to question 12," because the computer program will automatically bring respondents to that place in the survey.

Each page of questions and responses ideally should fit on a computer screen and require minimal scrolling. Items farther down on a computer screen may be missed if too much scrolling is required or if directions do not clearly tell the respondent to page down to see additional items. Lengthy online surveys should periodically give some visual information about how much of the questionnaire respondents have completed already and what they have yet to go. Unlike printed surveys, respondents cannot gauge how long the survey is or how much time they will need to complete it. Providing this information and estimated times as the survey is completed is helpful.

Various formats for entering data are available for designing computer-assisted surveys, including radio buttons, check boxes, and drop-down menus. *Radio buttons* are round dots, often appearing raised, which, when pushed by the on-screen arrow with a click of the mouse, change color and look pushed in. These pop back out if you select another answer, thus allowing only one answer for the question. These are best for mutually exclusive responses. *Check boxes* are better suited for questions that allow more than one answer, as in "Check all that apply." A click of the mouse on the box creates a check mark that remains, even if another box is checked. *Drop-down menus* are ideal for long lists of possible responses. Rather than trying to display all the options for an item on a screen that cannot contain them all, use a drop-down menu format, which involves clicking on the response box that opens to show the many options. For example, if you ask what U.S. state a person lives in, it's not good to list all 50 on the computer screen. Instead, use a drop-down box that displays them only when the respondent clicks on an arrow or symbol attached to the box.

Many programs allow the questionnaire designer to build in checks for data accuracy as responses are entered. For example, if you are asked to submit your birth year in four digits and only two are entered, the program should prompt you to reenter a four-digit number. Similarly, it's good to remind respondents that they failed to provide an answer for an item, thereby eliminating missing values. Prompts for missing answers or inaccurate entry of answers should be specific and direct the respondents clearly and politely in what they need to do. It's best to place the prompt near the missing or erroneous response. Remember, ethical issues also apply to online research, and respondents should have the option of not completing some items. Depending on the length of the survey and its focus, respondents should be able to complete the survey at a later time by following simple instructions. However, providing this option may result in incomplete surveys, when respondents forget to return to complete the questionnaire. E-mail reminders can help increase the response rates.

Not everyone is familiar with computers, so directions must clearly explain how to complete a survey online, for example, how to go through items on a screen by scrolling, where to click on a drop-down box, or to click or double-click the mouse button for other items. The opening page of the online survey should display a welcoming message and lead into a set of opening questions that are easy to complete and engaging enough to entice the respondent to continue with the survey.

Buttons or symbols that need to be clicked to continue or finish each page of the survey should be clearly marked and highlighted with colors that make them visible. Too many colors, though, can create eye fatigue if it is harsh or glaring, as well as confusion. Keep in mind how the creative features of computers such as colors, unusual fonts, graphics, sounds, videos, and other multimedia features can be misused. Not only do they slow down the process of loading and scrolling through the survey, especially for respondents with poor computer connections, but also these features may not always work with different kinds of computers, tablets, browsers, smartphones, and other multimedia equipment. As with all surveys, it is essential to pretest items and formats and to pilot-test the entire survey with different kinds of users and computer platforms.

Contact phone numbers, e-mail addresses, and help buttons should be provided to assist respondents who have concerns or questions while completing the survey. Issues of privacy must be addressed and displayed for respondents to read. Depending on your questions and whose computers are hosting the online survey, passwords might be needed for specifically chosen samples, and methods for encrypting transmission of responses might be necessary. Using passwords or some other codes also prevents people from responding more than once to the survey.

When deciding between paper and online questionnaires, remember that access to computers and limited computer abilities may affect the response rates and have a negative impact on the sampling design. Differences in computer usage can vary by sex, income, and education. Poor reading ability and language fluency, visual impairment, physical disabilities, and color blindness also may restrict the use of computer-based surveys, affect the sampling, and reduce the validity and reliability of the results. As with each research method or type of survey research discussed in this book, there are trade-offs; the pros and cons must be evaluated in the context of your research goals, respondents, and available funds and time.

PILOT-TESTING THE QUESTIONNAIRE

At this point, you have a completed draft of a questionnaire. Perhaps you have pre-tested some individual items, but now it's time to try out the entire survey. The best way of assessing whether the questionnaire flows, the instructions are adequate, the wording of the items and format are clear, and the survey takes a reasonable time to complete is to *pilot-test* it—first with yourself and then with others. Like any writing, the first version is a draft, and it only gets better with revisions.

Give the questionnaire to people similar to those who will make up the sample you want to study. These people, however, should not be part of your final sample because they have already seen the questionnaire, and having them take the survey a second time could bias the results. Distribute the survey or conduct the interview with all the same procedures you intend to use in the actual data collection phase. Remember, you are also pilot-testing the instructions and the procedures for distributing the survey.

When questionnaires are returned, read over the responses to the items to see if there is any confusion by looking for incorrect answers or marks left on the page by the respondents such as question marks or other annotations, for items consistently answered incorrectly or skipped, and for multiple responses that were selected when only one was expected.

You can also arrange to discuss the survey instrument with the respondents, asking them as they read each item to discuss what it means to them. Or you can wait until after they complete the survey and interview them in a focus group or individually about it. Encouraging them to say what they found confusing, how they reacted to the format and questions, and what they felt was missing will help you develop a final version. Some people give instructions to those participating in a pilot test to write comments on the questionnaire as they take it, although this makes it less like the way the final sample will answer the survey.

CODING QUESTIONNAIRES

When your surveys are completed and returned, getting them ready for data analysis is the next step. Transferring information on a questionnaire to a computer program (such as Microsoft Excel or the Statistical Package for the Social Sciences [SPSS]) is accomplished by assigning a value to each response category of a question, that is, coding the variables. Of course, if you created an online survey, coding is usually done automatically, and you just need to follow the hosting site's directions to download the responses to your statistical program or to perform some online statistical analyses. Even so, you still need to create questions and response formats that are appropriate for coding even by these online survey companies.

Although many computer programs accept text (alphanumeric) answers for nominal and ordinal variables, it is much more efficient to enter a numeral that stands in for the answer. In addition, when using words instead of numbers, the entered text must be exactly the same for each response: Typing in "anthropology" for a major is not the same as "anthro" or a misspelled "anthrapology." These entries would be tabulated as different majors. It takes fewer keystrokes to enter numbers, which are required when calculating certain statistics for many ordinal and all interval/ratio measures.

For example, if a question is how strongly respondents agree or disagree with a statement, it's much easier to enter a single-digit number such as "5" than to type out "strongly disagree" or even to enter "SD." This is why it is recommended that a number be assigned to each response. What helps greatly is to design your questionnaire so that every response to closed-ended items is already numbered, or precoded. Write the question with letters, for instance, "(a) Female (b) Male," or use numbers for the choices, such as "(1) Female (2) Male."

If the responses are based on interval/ratio measures, then the answer itself is its own coding. If you ask respondents to enter their age, the number they give is the number you enter. No coding is necessary. However, if you ask for height and some respond in inches (68") and others give you feet and inches (5'8"), you must convert them into the same units for comparison.

In addition, a *codebook* is needed to provide a detailed list of questionnaire items with their complete wording, the names of all variables (sometimes abbreviated, for example, PAEDUC for father's education as in the example in Box 4.3), the relevant codes for each response categories (such as 1 = female, 5 = strongly agree, and so on), the location of the code in the data file (usually given in terms of which column it can be found in on the spreadsheet, like Col. 112), and other guidelines for skipping responses or dealing with missing answers. Not only does a codebook serve as an aid to those doing the data entry, but it also is a reference guide for others using your data set later on.

Most programs (like SPSS or Excel) use a spreadsheet-style format for entering data into a file. Each row represents a unit of analysis, typically a respondent, but it could also be a school, a newspaper, a business, depending on what the study is about. Each column is a question, and what you put into each cell (the space where a row and column intersect) is the answer (the value) provided by the respondent.

Be aware, however, of a special case when dealing with a checklist of items for which respondents can select all that apply. Each item in the list essentially becomes a separate question. If you ask, "Which of the following activities did you engage in last week? (select all that apply): (1) playing a video game, (2) eating out in a restaurant, (3) doing aerobics, etc.," you will be coding each item separately. Remember, you cannot enter more than one answer in a cell. In essence, you are asking a series of yes/no questions, so you might enter a 2 if respondents watched TV and a 1 if they did not. Then you move on to the next item, entering 2 if they ate in a restaurant, 1 if they did not, and so on. Some researchers prefer using 1 for yes and 0 for no. Regardless of whether 0 and 1 or 1 and 2 coding is used, many suggest using the higher number for those who check the item "yes" and the lower number for "no." Each item on the checklist then becomes a separate variable and a separate column in your database.

Precoding the questionnaire and creating a codebook will result in fewer errors during data entry. If the coders have to remember whether male is 1 or 2, it is likely that some responses will be entered incorrectly. When the item is open-ended, more work and training are required in order to develop a set of coding categories. Issues of intercoder reliability must be considered, so that consistency among different coders is assured.

Content analysis is necessary for open-ended responses using a list of categories to summarize answers. For example, when a question asks people to describe their favorite leisure time activities, a coder needs to know what to do with such responses as "play sports," "watch TV," "read books," and "go to movies." Researchers must decide whether to code each possible answer with a different number (1 for sports, 2 for TV, 3 for reading books, etc.) or to create fewer categories by coding similar types of activities with the same value, such as 1 for active participation activities,

BOX 4.3

EXAMPLES FROM THE GENERAL SOCIAL SURVEY (GSS) CODEBOOK

PAEDUC
Highest year school completed, father

RESPONDENT'S FATHER'S (SUBSTITUTE FATHER'S) DEGREE

Response	Punch	Col: 106
Less than high school	0	
High school	1	
Associate/junior college	2	
Bachelor's	3	
Graduate	4	
Not applicable (no father/father substitute)	7	
Don't know	8	
No answer	9	

26. In what state or foreign country were you living when you were 16 years old?

REFER TO REGION CODES BELOW AND ENTER CODE NUMBERS IN BOX.

Col. 112
1 New England = Maine, Vermont, New Hampshire, Massachusetts, Connecticut, Rhode Island
2 Middle Atlantic = New York, New Jersey, Pennsylvania
3 East North Central = Wisconsin, Illinois, Indiana, Michigan, Ohio
4 West North Central = Minnesota, Iowa, Missouri, North Dakota, South Dakota, Nebraska, Kansas
5 South Atlantic = Delaware, Maryland, West Virginia, Virginia, North Carolina, South Carolina, Georgia, Florida, District of Columbia
6 East South Central = Kentucky, Tennessee, Alabama, Mississippi
7 West South Central = Arkansas, Oklahoma, Louisiana, Texas
8 Mountain = Montana, Idaho, Wyoming, Nevada, Utah, Colorado, Arizona, New Mexico
9 Pacific = Washington, Oregon, California, Alaska, Hawaii
0 Foreign

2 for passive/watching ones, or 1 for sports, 2 for entertainment, 3 for educational, and so on. Organizing responses using categories constructed before the questionnaires are returned is helpful, especially if these variables have been analyzed in other studies. Asking a native-born American to name the state where she or he

was born can lead to easy coding with standard regional breakdowns, such as East, West, North, and South. Consulting other studies, such as the General Social Survey (GSS), can provide other ways of organizing states into categories. On the other hand, you can wait until the questionnaires have been completed and then read the responses to help in the creation of meaningful categories. It's likely that you did not anticipate all possible answers, and by reading them through, you can develop more relevant coding schemes. Of course, you can combine both methods and have some prearranged categories and add to them as you discover new responses that don't fit those categories. For more details about doing content analysis, consult books on qualitative research methods and content analysis (such as Berg and Lune 2011 and Weber 1990).

Missing Answers

Inevitably, many respondents leave some questions blank because they missed them, refused to answer the question, didn't feel it applied to them, or didn't know how to answer the question. There are several ways of dealing with missing values. One way is to leave the response blank, and the computer program will treat it as simply missing and not use it to calculate any statistics. Or you can code the missing answer with a number not likely to be a real answer, such as −9 or 0 or 99. At some point, you have to instruct the computer reading the data that the code entered is not to be used in the calculation of any statistics. You can imagine what would happen if you were trying to figure out the average age of your respondents and you include −99 or 0 as answers! By coding a missing answer instead of leaving it blank, you can later analyze which people left the question out by having the computer search for all answers coded −9 and seeing if they are mostly, for example, males or females. You won't be able to tell, however, whether they left the question blank because they didn't see it or because they refused to answer it.

Sometimes you do know the reason for a missing value. People were asked to skip some questions through filtering and were branched elsewhere. You can code these absent answers as "not applicable" and assign a number to that response. There is a difference between those who skipped over the question because it didn't apply and those for whom it did apply but who didn't answer the question. Some researchers code "don't know" responses with a number and treat them as missing for later data analysis or use them to see which respondents were less likely to know the answers. In short, you might want to create a coding scheme to distinguish those people who simply leave it out by mistake, people who refused to answer the question, those who didn't know the answer, and those for whom it wasn't applicable.

ETHICAL CONCERNS IN QUESTIONNAIRE DESIGN

Given what you have learned about writing questionnaires, you can see there is some potential in designing surveys for intentionally manipulating the phrasing and order of questions to create biased results. Consider political opinion polls, which have a notorious reputation especially in highly contested elections. Each side will claim how inaccurate they are—that is, if they are behind in the polls! The next chapter discusses concerns about sampling, but the reliability and validity of data from surveys and polls are also strongly affected by the design of surveys, in particular, the ordering of questions and the way issues are framed and worded. Altering the wording, ordering of items, and the sampling of respondents to achieve biased outcomes introduces serious ethical concerns for the researcher.

One extreme unethical technique is the use of push polls. *Push polls* are typically used as a form of negative political campaigning to raise potentially harmful concerns about candidates. In some cases, businesses test new products by raising issues about the competition. For example, a poll might begin by stating political candidate A's opinions about abortion or immigration as a way of emphasizing a controversial wedge issue for the voters, reminding them about that politician's views, and pushing them away from that candidate toward their own. Often, the statement introduced in the push poll is not even accurate, as in the case of creating a negative impression by quoting a false rumor and asking people's opinion about it. Typically, these kinds of "surveys" are short, rarely ask demographic questions, and avoid other topics. They are essentially a form of political telemarketing solely designed to create false and negative impressions of the competition.

Yet, one does not have to be so blatantly unethical to skew a set of questions for a survey. Any of the tips presented in this chapter can be subtly manipulated to achieve desired or biased results. For example, a poll conducted many years ago found that about 25 percent of the respondents felt that too little money was being spent on welfare, while 65 percent said too little was being spent on assistance to the poor (quoted in Menand 2004). Obviously, the word *welfare* generated less support and is a loaded word. Other polls show that attitudes toward gay rights issues also depend on whether the question is asking about gay marriage, same-sex unions, or domestic partner benefits. Knowing this, a researcher could easily alter the results by phrasing items that lead to particularly favorable results for or against an issue or political candidate.

Ordering items on a survey in different ways can also lead to alternative outcomes, although some researchers feel it is more pronounced in interviews or online surveys than in self-administered questionnaires where respondents can more easily go back and change answers after reading ahead. Some researchers are concerned about the ethics of reporting results that may be affected by question order or wording, so they test various ways of phrasing items and try ordering them differently on parallel forms of a questionnaire. Consider an example from a 2008 Pew Research poll after Barack Obama was elected president.

In a survey of Americans that asked "whether Republican leaders should work with Obama or stand up to him on important issues" and "whether Democratic leaders should work with Republican leaders or stand up to them on important issues" (Pew Research Center 2011a), different responses resulted depending on which question came first. When asked first what Democratic leaders should do, 82 percent of people said they should work with Republican leaders. But if asked what Democratic leaders should do as the second question (after first asking about what Republicans should do), 71 percent responded that they should work with Republicans. That's 11 percent fewer people responding that they should work together. Similarly, when asked first what Republican leaders should to, 66 percent said they should work with Obama, compared to 81 percent who answered that way when the question came second. So which is it? Do 81 percent or 66 percent of the people believe Republicans should work with Obama? Do 82 percent or 71 percent feel that Democrats should cooperate with Republicans? This illustrates how context can make a difference in responses people give on surveys.

The ethics of questionnaire construction require researchers to eschew knowingly manipulating the order of items, the wording of questions, or the selection of a sample to achieve a desired outcome. Researchers are also obliged not to report results from questionnaire items that are discovered to be faulty after the data have been collected. Pretesting items and pilot-testing surveys are good steps in preventing inadvertent bias or intentional manipulation that could lead to the unethical use of research findings.

FINDING RESPONDENTS

When your questionnaire is ready for distribution, you need to find people to take it. No matter how good your survey is, poorly chosen respondents can seriously affect your results. Hence, it is crucial to use the best possible methods for selecting a sample of participants. Although you are very likely to have had in mind the kinds of people who will take your questionnaire in order to have written your hypotheses and constructed the items, finding such respondents requires many important considerations. The next chapter describes the strengths and weaknesses of various ways of generating a good sample for your study.

REVIEW: WHAT DO THESE KEY TERMS MEAN?

Anonymous versus confidential	Forced choice	Open-ended and closed-ended items
Attitudes	Instructions	Probe
Behavior	Intensity	Pilot test
Branching or skip pattern	Interview schedule	Ordering of items
Codebook	Leading questions	Push polls
Coding and precoding	Likert scales	Ranking versus rating
Content analysis	Loaded questions	Recoding
Demographics	Mutually exclusive and exhaustive	Response bias
Filtering	Online surveys	Social desirability

TEST YOURSELF

Consider these items from a questionnaire. Describe the main problems with the wording of these items and how you can improve the questions.

1. How often in the past week have you listened to news on the radio?

    a. 1 to 2 times

    b. 3 to 5 times

    c. 5 to 7 times

2. I enjoy watching sitcoms and drama shows on a regular basis.

    Strongly agree   Agree   Disagree   Strongly disagree

3. a. Are you currently enrolled in a science class?   Yes   No

    b. Which field of science is it (chemistry, physics, etc.)? _____

4. A recent analysis by a scientist from the Environmental Protection Agency concluded that global warming is affecting the sea level in countries that border the oceans. What is your opinion?

    Yes   No

INTERPRET: WHAT DO THESE REAL EXAMPLES TELL US?

1. Consider these questionnaire items from the 2010 General Social Survey (GSS), the leading national survey focused on the structure and development of American society and trends in its population's demographics, behaviors, and attitudes (see http://publicdata.norc.org/GSS/DOCUMENTS/OTHR/Ballot1_AREA_English .pdf).

    • Discuss the strengths and weaknesses of the wording and suggest alternative formats.

    • How would these items be written differently if it were a self-administered questionnaire instead of a face-to-face interview, or could they stay the same?

    a. Are we spending too much money, too little, or about the right amount of money on the environment?

    b. People have frequently noted that scientific research has produced benefits and harmful results. Would you say that, on balance, the benefits of scientific research have outweighed the harmful results, or have the harmful results of scientific research been greater than its benefits?

    c. Of all the telephone calls that you or your family receives, are (a) all or almost all calls received on cell phones, (b) some received on cell phones and some on regular phones, (c) very few or none received on cell phones, (d) don't know, or (e) refuse to answer.

    d. How much do you agree or disagree with each of these statements? (1 = agree strongly, 2 = agree, 3 = neither agree nor disagree, 4 = disagree, 5 = disagree strongly, 6 = can't choose)

    (1) It is just too difficult for someone like me to do much about the environment.

    (2) Many of the claims about environmental threats are exaggerated.

2. Figure 4.7 shows two pages from the General Social Survey (see Smith et al. 2011). This is a face-to-face interview survey. Notice how complicated a simple question about parents' occupation can become (in this case a father; information about mothers is asked in the same way later on).

    a. Discuss the branching and contingency question format and instructions to the people conducting the interviews.

    b. What other formats are used?

    c. Discuss the strengths and weaknesses of the wording and suggest alternative formats.

    d. How important is it to train interviewers?

Figure 4.7   Sample Pages from the General Social Survey

7. Were you living with both your own mother and father around the time you were 16?
   (IF NO: With whom were you living around that time?)
   IF RESPONDENT MARRIED OR LEFT HOME BY AGE 16, PROBE FOR BEFORE THAT.

   Both own mother and father..**(GO TO Q-8)** 1

   Father and stepmother...**(ASK A)** 2

   Mother and stepfather...**(ASK A)** 3

   Father–no mother or stepmother ....**(ASK A)** 4

   Mother–no father or stepfather...................................**(ASK A)** 5

   Some other *male relative* **(NO FEMALE HEAD)**
   (SPECIFY AND ASK A) 6

   Some other *female relative* **(NO MALE HEAD)**
   (SPECIFY AND ASK A) 7

   Other arrangement with *both* male and female *relatives* (e.g., aunt and uncle,
   grandparents)..**(ASK A)** 8

   Other **(SPECIFY AND ASK A)** 0

   A. IF NOT LIVING WITH BOTH OWN MOTHER AND FATHER:
   What happened?

   One or both parents died 1

   Parents divorced or separated...2

   Father absent in armed forces..3

   One or both parents in institution..4

   Other **(SPECIFY)**...5

   Don't know..8

IQ-3 INTERVIEWER: CHECK Q.7 IS A FATHER, STEPFATHER, OR OTHER
FATHER SUBSTITUTE SPECIFIED?
YES, FATHER OR SUBSTITUTE IS SPECIFIED
(CODES 1, 2, 3, 4, 6, 8)
(ASK: QS. 8, 9, & 10 FOR THAT PERSON)................................1

NO FATHER, STEPFATHER, OR OTHER MALE IS SPECIFIED
(CODES 5 or 7)
(SKIP TO IQ-4)...2

Figure 4.7   (continued)

---

8. A. What kind of work did your (father/FATHER SUBSTITUTE) usually do while you were growing up? That is, what was his job called?
OCCUPATION: _____

_____

B. **IF NOT ALREADY ANSWERED, ASK:** What did he actually do in that job? Tell me, what were some of his main duties?

_____

_____

_____

C. What kind of place did he work for?
INDUSTRY: _____

_____

D. **IF NOT ALREADY ANSWERED, ASK:** What did they (make/do)?

_____

_____

_____

_____

E. **IF ALREADY ANSWERED, CODE WITHOUT ASKING:** Was he self-employed, or did he work for someone else?

Self-employed..1
Someone else...2
Don't know 8

GSS 1998 #6
page 6

CONSULT: WHAT COULD BE DONE?

Figure 4.8 is an example of a poorly designed questionnaire. You are asked to review the survey and provide expert advice in making it better.

1. How many problems can you find with it?
2. How would you correct the errors and rewrite the questionnaire?

DECIDE: WHAT DO YOU DO NEXT?

For your study on how people develop and maintain friendships, as well as the differences and similarities among diverse people, respond to the following items:

1. Use the hypotheses or research questions that you wrote in Chapter 3 and/or develop some new ones. Suggest ways to measure the variables in them.
2. Write five attitude questions, five behavior questions, and five demographic questions that could serve as a good start for a much longer questionnaire. These items should measure some of the variables used in your hypotheses.
3. Work with another person or two and combine your questions to form a questionnaire of a reasonable length. If you are engaged in a research project, write the questionnaire you plan to use for the study.
4. Design a codebook for your questionnaire to explain how each variable is measured and coded.

Figure 4.8   A Survey of Questionable Merit

Attitudes about Cafeteria Food Questionnaire

This is a survey assessing your opinions about the food in the cafeteria, in order to improve the quality of the service and the food. Please place a check next to the appropriate answer. Your honesty in answering the questionnaire is appreciated. Your responses are confidential.

How often do you eat in the cafeteria? _____

Are you male or female? _____   How old are you? _____

Did you eat in the cafeteria yesterday?   Yes ___   No ___

Did you eat breakfast, lunch, or dinner? (Circle correct answer.)

Which of the following did you eat?

_____ cereal   _____ eggs   ____fruit

_____ vegetarian meal   _____ hot meal   ____dessert

Do you like to eat ham and eggs for breakfast? _____

What is your opinion about the new policy concerning hot meals?

Strongly agree   Agree   Disagree   Strongly disagree

Please rate the quality of food on a scale from 1 to 10 _____

What is your height?   Under 5' ____   5'–5'5"____   5'5"–6' ____   over 6' ____

Some people feel that the quality of food is just one part of what is going on in the institution and is part of a larger problem about quality of life here. What is your opinion about the organization as a place to be?

Strongly agree   Agree   Disagree   Strongly disagree

A recent analysis by a nutritionist from the Department of Health concluded that the food served in the cafeteria is some of the healthiest she has seen in some time. Do you think she is wrong?

            Yes   No

Why do you think she is not wrong?

Any other comments?

Thank you for completing this survey. Please return it to Mail Slot 999.

5

SAMPLING

Everyone takes surveys. Whoever makes a statement about human behavior has engaged in a survey of some sort.

—Andrew Greeley, sociologist

LEARNING GOALS

This chapter explains random probability sampling and describes other methods for obtaining samples. You will also learn about longitudinal and cross-sectional research designs. By the end of the chapter, you should be able to distinguish several types of probability and nonprobability sampling, describe various kinds of longitudinal research designs, and explain the idea of sampling error.

Ask your friends whether they have ever been surveyed by the Nielsen company about what television shows they watch. Every week the ratings come out with a listing of the most popular shows. Yet, so few people seem to know anyone who was ever asked by a pollster what they watched on TV that week. It is probably not surprising, given that only a few thousand people or so were actually surveyed. How can such a small sample of people accurately represent the entire nation?

Each election year, public opinion polls are routinely published with the latest voter preferences, and immediately the losing side claims the polls are not accurate. After all, only 1,000 people were surveyed in the typical Gallup Poll, and no one we know was asked. How then do we explain why the final presidential election results have almost always been within a few percentage points (the margin of error) of the last public opinion poll taken before the election when such a small sample of voters was measured? Typically, polls report the margin of error, and this number applies to each candidate's percentage, not to the spread between the candidates. So, if candidate A polls 45 percent and candidate B gets 49 percent and the margin of error is 4 percent, then the true results for candidate A are anywhere between 41 and 49 percent, and between 45 and 53 percent for candidate B.

Polling people for their opinions is a common practice in contemporary society. But not all polls are done accurately, and their reputations can suffer greatly. For example, your friends on Facebook may ask you to vote on some topic or the local news channel may ask viewers to text their responses to a question about some controversial issue. One of the most visible is CNN online, which reports the opinions of hundreds of thousands of people responding to a "Quick Vote" survey available on its website. You would think these findings are valid, given the large numbers of people who answer. However, even though the number of responses to the CNN surveys is substantially greater than the number reporting their television viewing habits to the Nielsen company, these polls are not scientifically accurate or generalizable. In fact, there is a small print disclaimer at the bottom of the CNN "Quick Vote" results, stating, "This is not a scientific poll." All these types of polls only reflect the opinions of those Internet users, cell phone texters, or TV news viewers who have chosen to participate and may not even represent the opinions of other users or viewers, let alone the public as a whole. Why this is so and how we can conduct a survey with a representative sample with minimal funding and time are the topics of this chapter.

SOME BASIC SAMPLING CONCEPTS

As mentioned in Chapter 1, the goals of research may be to describe, explain, explore, or predict characteristics of a population. A *population* is the total collection of units or elements you want to analyze. Whether the units you are talking about are a country's citizens, schools, editorials in newspapers, or local businesses, when the population is small enough, you can easily survey every element of the population. Of course, this means you have to be able to define clearly what is part of the population. For example, if you want to study the population of all students on a campus, you need to first determine who is a student: Do you include full-time and part-time students, those on leave of absence but still officially registered as students, or those studying abroad this year? And once you determine the actual population, where do you find a complete and accurate list of them? We know that a population survey of all Americans (that is, the census) misses quite a few people.

BOX 5.1
POLLING SAMPLES AND ACTUAL RESULTS

Look at the final Gallup Poll results for the winning candidate in presidential elections since the 1960s. These are the polls closest to Election Day and include results of likely voters with an allocation of undecided votes. The actual results are almost always within the usual (2 or 3 percent) margin of error.

	GALLUP POLL		
YEAR	FINAL ESTIMATE	ACTUAL RESULTS	DIFFERENCE
1960: Kennedy	50.5 percent	50.1 percent	0.4 percent
1964: Johnson	64 percent	61.3 percent	2.7 percent
1968: Nixon	43 percent	43.5 percent	−0.5 percent
1972: Nixon	62 percent	61.8 percent	0.2 percent
1976: Carter	48 percent	50.1 percent	−2.1 percent
1980: Reagan	47 percent	50.8 percent	−3.8 percent
1984: Reagan	59 percent	59.2 percent	−0.2 percent
1988: G. Bush	56 percent	53.0 percent	3.0 percent
1992: Clinton	49 percent	43.3 percent	5.7 percent
1996: Clinton	52 percent	49.2 percent	2.8 percent
2000: G. W. Bush	48 percent	47.9 percent	0.1 percent
2004: G. W. Bush	49 percent	50.7 percent	−1.7 percent
2008: Obama	55 percent	53.0 percent	2.0 percent
2012: Obama	49 percent	51.1 percent	−2.1 percent

Adapted from Gallup Polls (2013).

The *unit of analysis* is the element about which you are observing and collecting data, such as a person responding to a political poll, a school, an editorial, or a local business. Sometimes, however, the element surveyed is merely a mechanism to collect information about some other unit of analysis you need for your research questions. For example, you might survey school principals (the elements selected from the population of all principals) in order to gather information about student-teacher ratios in their schools. The unit of analysis is now the school and its student-teacher ratio, but the element selected randomly was the principal.

Let's focus on a group in which the unit of analysis is also the element selected from the population. If you want to know how everyone at your workplace feels about a new health benefits policy, and the number of employees is small enough, go survey each person. Then you would have information about the variables in this population of workers; these findings are usually called the population *parameters*. Or consider when a professor wants to find out what every student in a class thinks about the course and the teacher. As is done in most classes, a course evaluation is completed by surveying every unit of the class population, assuming no one is absent that day. The resulting data are the population parameters.

However, the population we are often interested in is usually larger than a class and, with limited time and money, we are unlikely to survey each and every one of the elements in a population. Therefore, in order to figure out what the population parameters are, we need to make inferences from a subset of the population; that is, we need to generate some *statistics* from a *sample* of people chosen to represent the entire population. Technically, the word *statistic* refers to the information we have about the variables in a sample. Population parameters are what we estimate or infer from sample statistics when we collect data from a sample because we were unable to do a full population study. In most cases, there will be a difference between the information (the statistics) we gather about the sample and what the true parameters of the population are. This difference is called the *sampling error*. As with political polls, for example, the errors are often stated in press releases as "plus or minus four percentage points."

In most cases, we need to generate a sample, but we still have to identify and count every element in the population in order to do accurate sampling. Sometimes employees, for example, are on leaves of absence. Should they be included in the list of every element of the population? Is this a study of both full-time and part-time workers? Of management and executives? In short, the researcher must first define the study population from which the sample will be selected and to which the resulting statistics will be generalized.

The best rule of thumb is that you can only talk about the population that is represented in the sample. If part-time students or fraternity/sorority members or off-campus students are not part of the population list (sometimes called the *sampling frame*) from which a sample is generated, then the resulting statistics cannot be generalized to all students at the university. The findings must be qualified as applicable only to full-time, on-campus, non-fraternity/sorority students. As discussed later, if the sample is obtained using what is called a nonprobability sampling method, then the information gathered can be used to describe, explain, or predict information about only those who completed the survey, that is, those who are part of the sample. You cannot say the data are meant to represent the opinions, behavior, or demographics of all students, or all young people, or any population of elements greater than those sampled.

✦ PROBABILITY SAMPLING

When we go for a medical examination and get a blood test, it's not necessary to take all our blood—a sample will do, thank you very much. Because blood is homogeneous throughout our veins, any vein will do, and as little as a drop or as much as a vial will suffice. Similarly, if the population we wanted to study were totally homogeneous, then any one element would do to represent the whole. But it is a very unlikely event that all elements are exactly the same on all characteristics in reality. Heterogeneity reigns and our population usually varies quite a bit in attitudes, behavior, and demographics. Therefore, we are obliged to take a sample of elements that represent what is in the population. For example, if the population we are interested in is composed of students with different majors, then our sample should reflect the range of majors in almost the same proportion that they exist on the campus. If 10 percent of the students major in sociology, then 10 percent of the sample should also be sociology majors.

In order to achieve this, we must use a sampling process that is likely to result in a representative sample. This will happen if each person has an equal chance of being chosen for the sample study. The proportion of people in the final sample who are

BOX 5.2
PROBABILITY THEORY, SAMPLING, AND M&Ms

M&Ms, my favorite candy, reports the following percentage of colors for their milk chocolate (plain) candies: 13 percent each of brown and red, 20 percent orange, 24 percent blue, 14 percent yellow, and 16 percent green. Imagine you have been invited by the company to dip a huge scoop into some large barrel of these candies at their factory. Let's say you picked out exactly 100 M&Ms. How many colors will you be tossing into your mouth? Based on their figures, you should have 13 brown, 14 yellow, 13 red, 20 orange, 16 green, and 24 blue ones in your scoop. And if you got only 50, you would have 6 or 7 brown, 7 yellow, 6 or 7 red, 10 orange, 8 green, and 12 blue ones.

So, purely in the interests of research, I went to the store and bought a package of M&Ms. The 1.69-ounce bag had 57 candies in it. Therefore, based on probability theory, it should contain 7.4 (13 percent of 57) brown ones, 8 yellow, 11.4 orange, and so on. Since it's unlikely that there will be 0.4 of a candy, let's round off the numbers and expect around 7 or 8 brown, 8 yellow, 7 or 8 red, 11 or 12 orange, 9 green, and 14 blue. Remember, one bag is a sample of all possible M&Ms (the entire population of candies is not in my little package!), and here's what I find: 14 orange, 9 red, 16 green, 7 brown, 9 yellow, and 2 blue. My colors are pretty close to the overall population distribution, with the exception of the blue and green ones. Perhaps there is some sampling error with a few of the colors, probably due to the small sample size.

BOX 5.2 CONTINUED

To make sure, I take another sample and buy a second package. (Well, somebody has to do this kind of research!) Remember, the proportion would be exactly what the company says it is, if I took all possible samples, but I'm only doing two. Guess what? In this one, out of 56 pieces, I have 6 red ones, 11 orange, 16 green, 4 blue, 13 yellow, and 6 brown ones. As random sample sizes increase, the sample statistics increasingly come closer to the actual population parameters, so let's combine the two packages to achieve a larger sample and see how close my sample comes to the expected distribution of colors.

	PACK #1 (N = 57)	PACK #2 (N = 56)	TOTAL (N = 113)	EXPECTED (N = 113)
Brown	7 (12.3%)	6 (10.7%)	13 (11.5%)	14.7 (13%)
Yellow	9 (15.8%)	13 (23.2%)	22 (19.5%)	15.8 (14%)
Red	9 (15.8%)	6 (10.7%)	15 (13.3%)	14.7 (13%)
Orange	14 (24.6%)	11 (19.6%)	25 (22.1%)	22.6 (20%)
Green	16 (28.1%)	16 (28.6%)	32 (28.3%)	18.1 (16%)
Blue	2 (3.5%)	4 (7.1%)	6 (5.3%)	27.1 (10%)

I could keep buying bags of candy and counting them by color until I had all the M&Ms ever made, but for a variety of dietary and financial reasons, I don't. By increasing the number of candies by the quantity in each additional bag, the percentage of each color becomes closer to the actual number produced by the company. The question becomes, Can I generalize to the entire population of M&Ms manufactured, based on any one sample of a single 1.69-ounce package? Theoretically, if I were to look at all possible samples, taking 57 candies at a time, the average percentage of all brown ones would be exactly 13 percent. Some packets would probably be 15 percent brown, others might be 11 percent, some would have extremely few brown ones, and a few would have extremely many brown candies. But when these sample percentages were averaged together, brown ones would be 13 percent of the total because, at that point, all possible samples of 57 would equal the total number of M&Ms made. If my sample sizes were larger—that is, if I bought those jumbo bags of candy—they would even more closely follow the percentages of the true population figures.

And if I plotted these sample percentages on a graph, they would form a normal curve. But since I wouldn't be able to get all these samples, it is a theoretical distribution. This is what is called the *central limit theorem,* which is discussed more in Chapter 6. It helps me decide whether my particular sample is representative of the true population information or whether I have a sample so extremely different that the probability of having gotten it by chance is less than 5 percent, the traditional cutoff point used in the social sciences to declare that this was not a typical sample or finding.

Our goal, then, is to obtain a sample that is representative of the population. If I asked someone to put together a bag of M&Ms for me, that person might select the colors he or she favored or might try to balance the mix, erroneously assuming that the company makes equal quantities of each color. It would, in either case, be a biased sample and unrepresentative of the true population. Because there is no reason to believe that factory workers are choosing the colors for each bag of M&Ms, we assume that the colors are selected randomly by the packaging machines, and therefore each bag (sample) should be close to the population figures, especially in the larger bags of candy.

BOX 5.2 CONTINUED

My total sample percentages were virtually the same percentage expected for red candies (13 percent), within two percentage points for the brown and orange ones, and pretty close to what was expected for the yellow M&Ms. But there were much larger deviations from the expected for the blue and green candies. Did I just happen to get unusual samples of blue and green ones by chance alone because I looked at so few packages of them? Larger samples would eventually produce numbers closer to the expected. The difference between what I observed and what I expected is called the margin of error or the *sampling error*. Like public opinion polls, my sampling distribution of percentages will not be exactly like the population's actual percentage breakdown of colors, but it will be within a few percentage points plus or minus the expected, especially when the sample is larger. I guess it's time to buy a bigger bag—for research purposes only, of course!

of different ethnicities, genders, sexual orientations, ages, and so on should look pretty close to their distribution in the population. If the population is 60 percent female, then the sample chosen should also be around 60 percent female. Unless you wanted to under- or oversample a subgroup (and this can be done by weighting or through disproportionate sampling techniques, both discussed in more advanced textbooks), probability sampling will assure approximate representation of the population.

When we can specify the (nonzero) probability of each element in the population being selected for membership in the sample, we have what is called a *probability sample*. The most common types include

- Simple random sampling
- Stratified random sampling
- Systematic random sampling
- Cluster or multistage sampling

Simple Random Sampling

Don't be fooled by the word "random," which seems to crop up regularly in everyday slang. It is a mistaken use of the word when someone tells you she stood on a street corner and "randomly" gave out questionnaires or walked up to people "randomly" in the mall or asked some "random" person to participate in a study. The researcher may have aimlessly wandered around finding people, but that doesn't mean every person chosen was randomly selected. Did every shopper, let alone citizen, have an equal chance of being designated a respondent? Not likely, because some people were working, others shopped earlier in the day, and still others were hanging out at the other end of the mall.

Therefore, in order to achieve a true *simple random sample,* you must be able to provide a complete list of all possible units in the population from which to choose a sample. Whether you are randomly selecting ads in a magazine, students on a campus, or clients of a social service agency, somehow you must first get a complete and accurate set of elements according to the criteria you decide. For example, if you want to generate a sample of employees where you work, then you must first decide what an employee is (full-time, has worked at least six months, is not on maternity leave, etc.) and be able to get a complete list of all who fit the criteria. This becomes the sampling frame from which a sample can be chosen.

All units in the sampling frame may be identified with a number, either computer-generated or done by hand, in order to use random sampling techniques. Each ad in a magazine could be assigned a number, every student already has an ID or mailbox number, each employee has a worker or social security number. Using either a table of random numbers (found in most statistics and methods books) or computer-generated ones (such as www.random.org/integers), the units of analysis are chosen (see Box 5.3). Or their names can be written on pieces of paper, placed in a box,

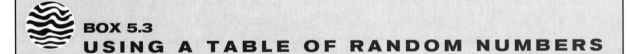

BOX 5.3
USING A TABLE OF RANDOM NUMBERS

Imagine this list of random numbers is part of a much larger table (available in most statistics books or generated with a computer program):

10570	15064	21121	15888	67710	04068	01509	64201	37714
74862	64821	63837	28633	78404	81599	35584	84476	50272
05235	92517	77904	43052	98066	48967	52408	02086	42821
78592	72486	89458	88743	81361	28906	29360	07461	24407

Although they look like five-digit numbers, these are just arbitrary spaces to make the table more readable. You can begin anywhere and decide to go in any direction, but not because you see some numbers you like. That would be biasing it. Once you choose a starting point (close your eyes and point to any number) and decide on a direction (for example, move to the left and then up when you get to the end of a line), keep going in that direction and choose the numbers that are within the range you need. Computer programs or Web-based programs (such as the one on www.random.org/integers) can also randomize phone numbers, student IDs, and so on for you, but if you are trying to get a relatively small sample from a small population, doing this by hand is not too difficult.

Let's say you had to choose 100 subjects from the 1,100 students on your college campus or employees at work. You find out from the mailroom that box numbers go from 100 to 1,200. A computer program can

BOX 5.3 CONTINUED

generate for you 100 numbers anywhere between and including 100 and 1,200. Or you can jump into the table of random numbers and begin looking at four-digit numbers. Even if you need only three-digit box numbers, you must use the number of digits that are represented in the highest box number you have.

Just to make it easy, let's start at the very first number (since these are randomly generated you can start anywhere in the table) and go to the right. The first four digits reading across are 1, 0, 5, and 7. So box number 1057 gets a questionnaire. Now go to the next four: 0, 1, 5, 0; remember the space is just used to make the table readable. Box number 0150 (that is, 150) gets a questionnaire. And continue on, in any direction, taking four digits at a time until you get your 100 box numbers, worker/student ID, phone numbers, or whichever numbers you are using. If you get a number already selected or out of range, in this case above 1,200, keep going.

However, because response rates for questionnaires are rarely above 50 percent the first time around, you may need to give out 200 questionnaires by using 200 box numbers just to get back 100. Theoretically, for every person who does not respond, your sample becomes less random. What if only the women tended to complete and return the questionnaire? It doesn't matter if the original 100 or 200 were randomly chosen; the final sample is truly random only if everyone selected responds. For many researchers, it is therefore important that the sample be representative of the population.

mixed well, and then, like a lottery, picked out at random until the desired number of respondents is selected. Some computer programs can also generate random samples just using the names, so there is no need to first assign numbers to each element in the sampling frame.

Telephone surveys employ probability sampling through random digit dialing techniques in which machines generate phone numbers within various area codes and then dial the numbers. Even those with unlisted numbers can be selected using this method. However, people without phones or who use only cell phones are not part of the sampling frame, and this possibility could bias the resulting sample by underrepresenting the poorest people. See Box 5.4 for a description of this technique as used by a national polling company.

Stratified Random Sampling

Let's say that you want a sample with comparison groups of equal size, even if they are not that way in the population. For example, you know that there are more female employees, but you need to be sure that half of your sample is female and the other half is male. What do you do now? Based on probability theory, your final random sample will come out looking like the population with more women in it, so somehow you have to ensure a 50-50 split. This is where *stratified random sampling* comes in. With this method, you disproportionately stratify (categorize) your

BOX 5.4
RANDOM DIGIT DIALING EXPLAINED

Here is a description of the sampling methods used by the Pew Research Center for the People and the Press, one of the leading national survey organizations:

The typical Pew Research Center for the People & the Press national survey selects a random digit sample of both landline and cell phone numbers in all 50 U.S. states and the District of Columbia. As the proportion of Americans who rely solely or mostly on cell phones for their telephone service continues to grow, sampling both landline and cell phone numbers helps to ensure that our surveys represent all adults who have access to either (only about 2% of households do not have access to any phone). We sample landline and cell phone numbers to yield a combined sample with approximately 60% of the interviews conducted by landline and 40% by cell phone. This ratio is based on an analysis that attempts to balance cost and fieldwork considerations as well as to improve the overall demographic composition of the sample (in terms of age, race/ethnicity and education). This ratio also ensures a minimum number of cell only respondents in each survey.

The design of the landline sample ensures representation of both listed and unlisted numbers (including those not yet listed) by using random digit dialing. This method uses random generation of the last two digits of telephone numbers selected on the basis of the area code, telephone exchange, and bank number. A bank is defined as 100 contiguous telephone numbers, for example 800-555-1200 to 800-555-1299. The telephone exchanges are selected to be proportionally stratified by county and by telephone exchange within the county. That is, the number of telephone numbers randomly sampled from within a given county is proportional to that county's share of telephone numbers in the U.S. Only banks of telephone numbers containing three or more listed residential numbers are selected.

The cell phone sample is drawn through systematic sampling from dedicated wireless banks of 100 contiguous numbers and shared service banks with no directory-listed landline numbers (to ensure that the cell phone sample does not include banks that are also included in the landline sample). The sample is designed to be representative both geographically and by large and small wireless carriers.

Both the landline and cell samples are released for interviewing in replicates, which are small random samples of the larger sample. Using replicates to control the release of telephone numbers ensures that the complete call procedures are followed for the entire sample. The use of replicates also ensures that the regional distribution of numbers called is appropriate. This also works to increase the representativeness of the sample.

When interviewers reach someone on a landline phone, they randomly ask half the sample if they could speak with "the youngest male, 18 years of age or older, who is now at home" and the other half of the sample to speak with "the youngest female, 18 years of age or older, who is now at home." If there is no person of the requested gender at home, interviewers ask to speak with the youngest adult of the opposite gender. This method of selecting respondents within each household improves participation among young people who are often more difficult to interview than older people because of their lifestyles.

Unlike a landline phone, a cell phone is assumed in Pew Research polls to be a personal device. Interviewers ask if the person who answers the cell phone is 18 years of age or older to determine if they are eligible to complete the survey.... This means that, for those in the cell sample, no effort is made to give other household members a chance to be interviewed. Although some people share cell phones, it is still uncertain whether the benefits of sampling among the users of a shared cell phone outweigh the disadvantages.

Source: Used by permission from The Pew Research Center for the People and the Press, http://people-press.org/methodology/sampling/random-digit-dialing-our-standard-method.

sample along the lines you want to analyze by establishing quotas for certain kinds of respondents. Take religion: you might want 20 percent Roman Catholic, 20 percent Protestant, 20 percent Jewish, 20 percent no religion, and 20 percent other religions. So you divide the sampling frame into these categories and within each category (stratum), you take a simple random sample using the methods described previously until you get the proportion of respondents you desire for each category.

Stratified sampling can also be a way to guarantee an exact proportionate representation of the population, even if simple random sampling might yield almost similar results. If you know that the population is 60 percent female and 40 percent male, you can stratify on gender and achieve a 60:40 ratio in your final sample, rather than leave it to chance. You could also stratify on a combination of traits, such as both sex and height, in which you want each 25 percent of your sample to be men 5'9" and under, men over 5'9", women 5'9" and under, and women over 5'9". Whatever strata you use, you must first be able to identify people who fit into these categories and then perform random sampling within the strata. That's the key part for a probability sample: Without randomizing respondents, you would have a nonprobability quota sample (discussed in the section on Nonprobability Sampling) and lose the ability to generalize the results to a larger population.

If you don't know how many people in different categories responded until they completed the demographic section of the questionnaire, you can use statistical weighting techniques during the data analysis phase to adjust for under- and over-represented groups. This is discussed in advanced statistics books and available in statistical computer programs like SPSS.

Systematic Random Sampling

Another way of generating a random sample, especially if there is a large number of population members, is to use systematic random sampling. This involves taking every nth element in the sampling frame until the total is reached. Imagine you need a sample of 100 and you have 5,000 people or any other elements such as magazine ads, schools, or TV characters to select from. If you divide 5,000 by 100, you get a sampling interval of 50. So you take every fiftieth person, ad, school, or whatever from the population list.

Let's use mailbox numbers on a college campus as an example of reaching each person in the population. Imagine these begin at number 100 and go up to 1,200, and you want a sample size of 100. You must begin randomly somewhere between and including mailbox number 100 and number 1,200. Since you have 1,100 mailbox numbers and need 100, your sampling interval is 11. Begin by getting a random number from 100 to 1,200 (either using a printed table of random numbers or a computer-generated one). Let's say you randomly select number 500 as your starting point. Mailbox 500 gets a questionnaire then box number 511, 522, 533, and so on, until you get 100, increasing in sampling intervals of 11. If you want to distribute 200 questionnaires instead, then divide 1,100 by 200 to get a sampling interval of 5.5; you would now choose every sixth mailbox (because there is no such thing as a half mailbox) after starting randomly, such as 500, 506, 512, 518, and so on, until you select 200 mailbox numbers. There is no need with systematic random sampling to generate 200 random numbers, only a random starting one.

The sample resulting from this systematic method is typically quite similar to a simple random sample, unless for some reason there is a pattern to the order of the population elements. For example, consider the case where stratified random sampling selects every tenth house on a city's streets. At some point you notice that every tenth house happens to be on a corner lot containing larger houses, probably with higher-income residents. The final sample could end up disproportionately composed of upper-class respondents. In this situation, a systematic random sample would not generate a representative group of people.

Multistage or Cluster Sampling

When even larger populations are used from which to select a sample, researchers use what is called a *multistage* or *cluster sampling* method. Large national polling agencies typically employ this method to generate samples. This method involves randomly selecting units beginning with larger clusters and moving to smaller ones at each stage. If, for example, you are interested in surveying Americans' attitudes toward a presidential candidate, you might begin by randomly choosing a certain number of states, either stratified by region or simply taken from the entire list of 50. Then, the next stage is to randomly select counties from within the selected states. The third stage involves randomly choosing cities within the counties now selected. The next stage is randomly generating streets, and the final stage is randomly choosing houses on those streets. At each stage, simple, stratified, or systematic random sampling can be done. This is how pollsters can get away with 1,000 people representing the entire population of the country and why it is unlikely you would know anyone selected for such a survey!

Although this technique is more useful when you have very large national samples, it's possible to do something like this at a big university. Make a list of residence halls and fraternity/sorority houses. The first stage randomly selects from this cluster. Perhaps if the residences are large enough, you can then choose a second cluster from floors, then randomly select rooms, and finally pick roommates at random. Of course, such a sample would not include off-campus students. But if your study is about residential life, then this random multistage method allows inferences to be made about the entire population of residential students, but not about all students at the university.

These probability sampling methods can be combined; perhaps a stratified systematic random sampling is useful for some studies, or stratified sampling can be used at the various stages in cluster sampling. The key point for any of these is that every element should have a specified (usually equal) chance of being selected into the final sample.

NONPROBABILITY SAMPLING

Sometimes it is just not practical, cost- or time-efficient, or necessary to do a true random sampling. However, it must always be kept in mind that, unless there is a random selection of units from the population, you cannot generalize to the entire population. With nonprobability methods—those for which every element does not have an equal chance of being selected for the study—you are limited to making conclusions about only those who have completed the survey.

There are several kinds of nonprobability sampling, including

- Convenience or accidental sampling
- Purposive or judgmental sampling
- Quota sampling
- Snowball sampling

Convenience or Accidental Sampling

Everyone at one point fills out some sort of questionnaire, whether it's a warranty card for a product recently purchased or as part of some major academic study in a class you took in college. Perhaps you have been stopped while shopping at a mall and asked to answer some questions. Maybe you wasted some money texting your vote for your favorite singer on *American Idol* or answered a customer survey that popped up after you visited an Internet website. While you may have felt like a good citizen in helping out, what you may not be aware of is that you have been part of a nonprobability and typically nonscientific *accidental* or *convenience sample,* one from which it would be difficult to generate any reliable information about a population of people, despite misleading statements by people using these data to imply that these are what *all* citizens believe or do.

All that can be reliably said is that the data resulting from these studies apply only to those who were available and attending that particular psychology class the day surveys were distributed, who had money to text a message on a mobile phone, who had access to the Internet and linked to that web page, or who were stopped on that day at that hour in that particular mall. To say anything about all college students, all listeners of a radio show, all Internet users, all shoppers, or for that matter all citizens in the country is impossible. But too often these data are inadvertently misused in just that way, and sometimes they are used unethically on purpose.

Convenience samples are based on whoever just happens to be available at a particular moment or accidentally walking by the person distributing the survey. Everyone does not have an equal chance of getting selected. While it certainly is more convenient to go to an intro economics class and get 100 students at once than it is to generate 100 random numbers and stuff mailboxes with questionnaires, the results have limited generalizability.

Volunteer samples, sometimes referred to as "opt-in," or self-selected, respondents, are also samples of convenience. Those who respond to a sign asking for research subjects, participate because of some incentive (a course grade or money), or hear about a survey on some website may be different kinds of people from those who do not even see those announcements. Again, all people do not have an equal chance to select themselves into these studies. And even if they did see calls for survey respondents, those who volunteer may be a very different type of person from those who don't volunteer, such as they need money, have extra time to participate, or are interested in the particular topic of the study.

Purposive or Judgmental Sampling

Sometimes there is a specific reason to choose a unique sample on purpose, because of some characteristics of the units of analysis. *Purposive* or *judgmental sampling* involves designating a group of people for selection because you know they have some traits you want to study. For example, you choose a particular sorority on campus for study because prior research has shown that it represented the feelings of most students on campus in the past. Many times subscribers to magazines get surveyed to find out more about the characteristics of their readers. Marketing researchers test products in a particular city because they have made the judgment that shoppers in that city represent a cross-section of potential buyers nationally. Although people in these kinds of samples are not randomly selected, prior research may have indicated that their patterns of buying were representative of the entire population and could be purposely chosen to make inferences about all potential buyers. In all these situations, every person does not have a specified chance of being selected for the study, only those who are available and purposely chosen.

Quota Sampling

Like stratified random sampling, researchers occasionally want to be sure that there is some representation in the final sample. An accidental sample might involve researchers stopping the first 100 people walking out of a movie theater to survey them about the film. Because it is not a random sample, the number of men and women leaving first may not be in the same proportion as they were in the theater. If you wanted to guarantee that half are male and half are female, a *quota sampling* would be done in which the first 50 males and the first 50 females coming out of the theater are surveyed. It is still an accidental sample, but one using a quota system. Breaking your sample into various strata can entail any number of categories, such as race/ethnicity, sex, age, and any other demographic characteristic used to screen respondents. Respondents are solicited until the number needed for each of the various criteria is met.

Snowball Sampling

There may be times a study needs respondents who are difficult to find or identify. What if you wanted to do a survey of gay men and lesbians? A stratified random sample or a nonprobability quota sample would require that you identify everyone's sexual orientation before getting your participants. This is a question people might not want to answer when stopped accidentally in a shopping mall! So what researchers do to get such a sample is first to identify a handful of gays and lesbians, perhaps through some personal contacts or organizations. Each person who volunteers then is asked to pass along a questionnaire to someone he or she knows who is also gay or lesbian, like a snowball rolling down the hill that becomes larger and larger as it picks up more snow. This is why this technique is called *snowball sampling*.

Let's say you wanted to study housekeepers. You survey a few chosen through various channels and then ask them to provide names of other housekeepers they know who would be willing to participate in your study. It helps if people are contacted first by their friends and then give permission to provide their names. You survey them and ask for a list of more names, and so on and on. Clearly, this is not random, and the final sample is made up of networks of people who tend to be somewhat similar in other characteristics (social class, race/ethnicity, etc.), since people tend to pass along questionnaires to those who are their friends. Studies of friendship demonstrate that people's friends are similar to themselves in many basic characteristics like educational level, race/ethnicity, marital status, and income.

CROSS-SECTIONAL AND LONGITUDINAL STUDIES

When a survey is given at one point in time and only once to a particular sample of respondents, it is referred to as a *cross-sectional* study. These are not ideal for uncovering causal relationships that require demonstration of a time sequence for the independent and dependent variables, but they are easier to do and require less time commitment than *longitudinal studies* in which samples of respondents are followed over lengthy periods. One kind of longitudinal study involves following the same people and surveying them at different points in time. This is called a *panel study,* but it has the risk that people cannot be found later on due to moving, death, changing names, or simply lack of interest. Let's say you randomly select a group of 100 graduates from the city's high schools, and then every five years, you contact them to see what they are doing with their lives. But, as it happens, not everyone can be found years later. What may have begun as a random sample, now due to attrition, ends up as a biased one and composed of volunteers who want to participate.

Another version of a longitudinal study is a *cohort study,* which involves sampling, at different periods of time, different people who are similar to the first sample, usually in age. Perhaps researchers surveyed 20-year-olds in 1990, then studied a different group of 30-year-olds in 2000, interviewed a sample of 40-year-olds in 2010, and plan to survey 50-year-olds in 2020. Even though the samples are not composed of the same people, they represent people who were born the same year (the same cohort) and have probably experienced the same major historical events and changes in the culture.

Like public opinion polls, some studies are interested in seeing how the same behaviors or attitudes change over time. Every year since 1975, the University of Michigan's *Monitoring the Future* study continues its survey of high school seniors' drug and alcohol use. This is a *trend* study in which different seniors are surveyed about the same behaviors across different years in order to track changes. These are not the same people followed over time, and they are not those born in the same period or cohort. Figure 5.1 shows the differences in cross-sectional and longitudinal samples.

SAMPLE SIZE

A common worry is how large a sample should be to have a reliable and valid study. Although mathematical models determining sample size are discussed in advanced statistics books (they involve "power analysis" and the calculation of confidence limits based on accuracy at, say, 95 percent or 99 percent confidence levels), there are some points to keep in mind for a beginning researcher. When a population is more homogeneous, fewer elements are required to get a representative sample.

Figure 5.1   Cross-Sectional versus Longitudinal Samples

	RESPONDENTS	TIMELINE
Cross-sectional	One set	One time
Longitudinal: panel	One set	Two or more time periods
Longitudinal: trends	Different people	Two or more time periods
Longitudinal: cohort	Different people, shared characteristics	Two or more time periods

The more heterogeneous a population is on a variety of characteristics (such as race, gender, age), the larger a sample is needed to reflect that diversity. Stratified random sampling can achieve representation with a smaller number than simple random sampling would require to ensure representation of the diverse population.

Sample size also depends on what is being studied. If you want to find out information about something that occurs less frequently in the population you are studying, such as marital life among 16- to 20-year-olds, then you will need a large sample to find enough respondents. On the other hand, if the behavior and attitudes you are interested in are much more likely to occur (like the use of cell phones) or large differences or relationships are expected, then a smaller sample size is sufficient.

One simple answer is the larger the sample size, the better. Sampling error is greater when making inferences to a large population from a small sample with a large standard deviation, especially if the population is diverse (recall the M&Ms example; also see Chapter 6 for more about sampling error and the concept of the *central limit theorem*). But remember, if you are using nonprobability techniques, it doesn't really matter whether 100,000 or 100 people text a message to voice their opinion or cast a vote; the results are limited to those who texted. Generalizing to the entire population is not possible, regardless of the sample size.

Another way of estimating sample size is to consider the kinds of analyses you plan to do. If you are interested in comparing subgroups, such as sex differences within 5 categories of ethnicity/race, then you will need enough people to fill 10 different categories—five ethnicities by two genders equals 10 categories: Asian/Pacific women, Asian/Pacific men, Native American men, Native American women, and so on. You are not going to be able to do much in the way of sophisticated statistical analysis if you do not have at least five people in each of those 10 categories. Already you need 50 respondents, assuming they fall equally into these categories. If you use stratified random sampling you might be able to achieve this, but it may take hundreds of people selected by simple random sampling to get at least five in each category.

Given the limited time and money most of us have to do a short survey, especially for class projects or theses, consider giving out 100 surveys using random sampling techniques with the hope that half will reply (after some reminders) in order to get a small sample size of 50 respondents. Usually 20 to 30 percent of people who receive questionnaires return them right away. Reminder mailings or calls can bring that percentage up to 50 percent or more. Response rates under 60 or 70 percent may compromise the integrity of the random sample.

Sometimes it is necessary to supplement randomly distributed questionnaires using nonprobability methods in order to increase sample size. However, you must be sure to code them in some way to see if the responses from this group are different from the randomly chosen ones. In any event, once you introduce nonprobability sampling, you are limited to making conclusions only about those who filled out and returned the survey. For most small surveys, this might be all you really need.

How you distribute your questionnaires can affect the quality of the sampling. Mailing them with return postage and envelopes requires some funding, but prepaid postage increases response rates. Giving them to people and asking them to send the surveys with their own postage results in fewer returned ones, especially for those on limited incomes. Dropping by in person and picking surveys up from respondents later might be a better way of getting them back if this is convenient to do. Online and e-mailed surveys are easier to distribute, but you are not likely to know who completed them. This type of distribution can also result in a nonrandom sample limited to those people with access to and comfort with using computers.

Following up with reminders (regular mail, e-mail, phone calls) one week after distribution, for example, and another reminder a week or 10 days later, increases response rates. But when names or identifying numbers are not used, you have to send reminders to everyone, including those who already returned their surveys. It is essential that the reminder notice thank those who have already sent their forms back.

Distributing questionnaires in large settings, such as a class, and waiting for them to be completed generates a much better response, since nearly everyone fills them out in such settings, but the nonprobability sample may not be ideal. In general, it is considered good if you get around 50 percent of questionnaires completed in the first phase, but don't be surprised if the response rate is much smaller. After several reminders, it is very good if your response rate is over 70 percent.

At this point in the research journey, you should have a clear set of research questions or hypotheses guiding the development of a clearly written valid and reliable questionnaire, a sample ready and eager to take your survey, and a method for distributing and collecting the forms. Now you eagerly await their completion and make preparations for coding the data for statistical analyses (see Chapter 4). The next four chapters discuss various ways of presenting results and analyzing your data.

REVIEW: WHAT DO THESE KEY TERMS MEAN?

Cohort studies

Convenience or
   accidental sampling

Cross-sectional design

Longitudinal design

Multistage or cluster sample

Nonprobability sampling

Panel studies

Population

Population parameter

Probability sampling

Purposive sampling

Quota sampling

Random digit dialing

Random sample

Sample size

Sample statistic

Sampling error

Sampling frame

Snowball sampling

Stratified random sample

Systematic random
   sample

Trend studies

Unit of analysis or element

Volunteers

TEST YOURSELF

For each of the following studies, state what method of sampling was used.

1. Researcher surveyed the first 25 women and the first 25 men who entered the cafeteria during lunch.
2. As part of their requirements for a course, all the students in "Introduction to Psychology" completed a questionnaire.
3. The researcher divided the number of people she wanted to survey by the total number of people in the population and then chose every nth person to get the survey.
4. Members of the local motorcycle club completed questionnaires and were then asked to give questionnaires to other people they knew who owned motorcycles, who were also asked to give some questionnaires to their motorcycle friends.

INTERPRET: WHAT DO THESE REAL EXAMPLES TELL US?

For the following examples from actual research, discuss what sampling strategy is used and what the limitations and strengths might be with each type of sample:

1. For a study of families' involvement in the criminal justice system and child welfare, researchers analyzed child protective case records: "A systematic random sample of 113 cases was selected by beginning with the first case [of 452] and selecting every 4th case thereafter" (Phillips et al. 2010: 546).

2. A study in Japan of health issues and the perception of trust and reciprocity in the workplace studied workers in companies "stratified into three categories according to the number of their employees ... [and] randomly selected 20 companies in each stratum" (Suzuki et al. 2010: 1368).

3. For a study on sexually risky behavior among gay and heterosexual adolescents, "A multistage cluster sampling design was employed to obtain a representative sample of adolescents enrolled in public high schools.... At the first stage of sampling, 63 of 299 schools with 100 or more students in grades 9 through 12 were randomly selected.... On average, 3 to 5 required classes (e.g. English, homeroom) per school were randomly selected at the second stage of sampling, yielding a total of 5,370 students, 4,159 of whom completed the survey (77% response rate)" (Blake et al. 2001: 940).

4. A survey of beliefs about family responsibilities, gender, and work roles "employed a randomly selected national sample. Two thousand thirty-three married people were first interviewed by telephone in fall 1980 and then interviewed again in fall 1983, 1988, and 1992.... Sample attrition is relatively high at 41 percent when the 1992 group is compared to the original group" (Zuo and Tang 2000: 32).

5. In comparing men's friendships between two occupations with different masculinity images, "a non-random quota sample of ninety-eight men (40 grade school teachers, 58 enlisted military personnel) was obtained" (Migliaccio 2009: 230).

6. An article on how race affects informal social encounters between people, especially those of the general category "Asian," reports that its sample of 64 Chinese Americans and Korean Americans was "located through the membership lists and referrals of a variety of churches, professional and social clubs, and college and university alumnae associations. The sample was expanded through 'snowballing'" (Kibria 2000: 80).

CONSULT: WHAT COULD BE DONE?

A study was completed at a university focusing on students' use of alcohol and drugs. Questionnaires were left in everyone's mailboxes and students were told to return them to a particular location by a certain date. Almost 25 percent of the students returned the surveys and the results were printed in the campus newspaper. The author of the article concluded that an extremely high percentage of students at the school drank or took drugs, despite anecdotal evidence to the contrary. You are asked to consult and explain more about the sampling.

1. What might be going on here?
2. What could you do to remedy any problems?
3. Describe two other sampling methods you would recommend instead and the strengths and weaknesses of each.

DECIDE: WHAT DO YOU DO NEXT?

For your study on how people develop and maintain friendships, as well as the differences and similarities among diverse people, respond to the following items:

1. Design a sampling strategy that involves a purposive method. Whom would you survey and why?
2. How would you go about getting a random sample of people in your local community?
3. Describe several ways of getting an accidental or convenience sample for this study.
4. In each case, list the strengths and weaknesses of the sampling design.
5. If you were to do a study of friendships as depicted in television sitcoms, and you wanted to do a content analysis of the shows themselves, describe how you would go about designing a random sample of TV shows.
6. If you are doing your own research project, select a sampling method, describe how you plan to survey respondents, and generate an actual list of respondents, mailbox numbers, or other relevant information.

PRESENTING DATA

Descriptive Statistics

Just think of how stupid the average person is, and then realize that half of them are even stupider!

—*George Carlin, comedian*

LEARNING GOALS

Understanding how to describe your findings with graphs, tables, and statistics is the focus of this chapter. By the end of the chapter, you should be able to decide how to use the mean, median, mode, and standard deviation when presenting data. You should also understand the concept of the normal curve and z-scores. In addition, you will learn the concepts of probability and statistical significance.

Figure 6.1 Statistical Decision Steps (also see Statistical Analysis Decision Tree in Appendix)

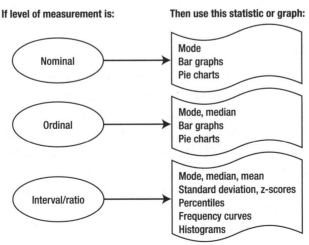

Finally comes the big moment: What did you discover from your surveys? You eagerly await the outcome after weeks, maybe months, of designing a questionnaire, developing a sampling strategy, distributing surveys, begging people to complete them, and coding and entering data into a computer program. Now what do you do? Where do you even begin? How are you going to make sense of all these responses to the questionnaires? Presenting data, analyzing relationships among the variables, and applying statistical tests to the findings are the focus of the next four chapters. This chapter discusses the first steps in describing the data and introduces some basic theoretical concepts about the normal curve and significance level.

PRESENTING UNIVARIATE DATA

If we want to describe, explain, explore, or predict some phenomenon, we must first be sure that the variables in the study are actually variables. Imagine an extreme situation in which the only people to have completed your survey were all women. The sex variable is now a constant and you could no longer use sex as a variable in explaining or predicting any other variables in the study. Hence, before any further data analysis is accomplished, it is important to do some *univariate* (one variable at a time) analysis and look at every item in the study to get a sense of the variability of responses. This is necessary in order to decide whether the variables can be used for additional statistical analysis.

In the process of doing a descriptive analysis of the variables, you also get a profile of the respondents with the demographic items and descriptive data for the other behaviors and attitudes measured. These may be all that are needed for a descriptive study but just the start for other kinds of research. There are several ways of presenting univariate information about the variables in your study, including frequency distributions, graphs, and statistical measures.

Frequency Tables

A *frequency table* or *distribution* shows how often each response (a *value*) was given by the respondents to each item (a *variable*). Frequency tables are especially useful when a variable has a limited number of values, such as with nominal or ordinal measures. It is less useful when an interval/ratio variable has many values: For example, when age varies from 18 to 89 in your study, the output could have over 70 rows of numbers, making the table all but unreadable.

The frequency for each value is listed in absolute raw numbers of occurrence and in percentages relative to the number of total responses. Percentages can be presented in terms of both the total number of questionnaires coded and the total number of those actually responding to the question (sometimes called the *valid percent*). A *percent* is the proportion of responses standardized on the basis of 100. "Per cent" means "per 100" from the Latin *per centum* for "by hundred."

A proportion is calculated by dividing the number of responses given for a particular value by the total number of responses for the variable; then that proportion (usually somewhere between 0.0 and 1.0) is multiplied by 100 to get the percent. Sometimes if occurrences in a population are small, such as crime rates, numbers are presented "per 1,000" or "per 10,000" instead of "per 100." For example, the U.S. Bureau of Justice Statistics reports that in 2010 there were 120 property crimes (such as auto thefts, burglaries) per 1,000 households. Note that the unit of analysis in this case is a household, not a person. This means that for every 1,000 households in the United States, 120 experienced some property crime in 2010. This would be the same as saying 12 for every 100 households, or simply 12 percent.

Let's review a frequency table, as presented by output from the SPSS program (Table 6.1). Three people of 154 in a survey of college students taking a research methods course did not report their political party affiliation. Therefore, 88 respondents said they were Democrats, representing 57.1 percent of the total number of people who completed the questionnaire (154), but 58.3 percent of those who actually answered the question (151); this is the valid percent. The cumulative percent is useful only for ordinal or interval/ratio measures since it requires that the values accumulate in some order.

We could conclude that there seems to be a variable here; it is not a constant. How do you decide if this is the case? What if 90 percent said they were Democrats, or 80 percent, or 75 percent? When is variability evident? There is no set rule: You

Table 6.1 Frequency Table, SPSS

		FREQUENCY	PERCENT	VALID PERCENT	CUMULATIVE PERCENT
	POLITICAL PARTY				
Valid	Democrats	88	57.1	58.3	58.3
	Republicans	23	14.9	15.2	73.5
	Independents	31	20.1	20.5	94.0
	Other	9	5.8	6.0	100.0
	Total	151	98.1	100.0	
Missing		3	1.9		
	Total	3	1.9		
Total		154	100.0		

have to decide if there are enough respondents in each of the categories (values) of the variable to do further data analysis. Clearly in this case, only nine people have another political identity than the three most common ones, and "other" would not be a useful category for further analyses, although the variable itself can still be used. Perhaps the "other" category should be combined with the "Independents" by recoding the data.

Charts and Graphs

In addition to a table of numbers, you can represent your findings visually with a graph or chart. If the variable has a limited number of discrete values, as with nominal or ordinal measures, then select a *bar graph* or *pie chart* to illustrate what you found. These graphic representations give a quick visual description of your variables. Figure 6.2a shows a pie chart for religion. Figure 6.2b shows the same information with a bar graph.

If your data are continuous or interval/ratio measures, histograms and frequency curves (sometimes called *frequency polygons*) are better ways of visually presenting univariate information. *Histograms* (Figure 6.3) are similar to bar charts, but the bars are adjacent and touching each other to indicate the continuous nature of the measure. Their width and height communicate the number of responses grouped within some interval. The intervals of the values for the variable are placed along the horizontal or *x-axis,* and the frequency range, in raw numbers or in percentages, is designated along the vertical or *y-axis* of the graph.

A frequency polygon (Figure 6.4) is the result of connecting the midpoints of each of the intervals (bars) in the histogram with a line. Other visual ways of displaying information, such as other line graphs, stem-and-leaf displays, box plots, and stacked bar graphs, are described in more advanced statistics books and online; these graphs are available in most statistical computer packages. Line graphs are especially useful for depicting changes over time.

The Normal Curve. One of the most important frequency polygons is the normal curve; it is at the core of most social science statistics and methodologies. Whether or not a variable is normally distributed in a population can affect the interpretations made about the results. By definition, a *normal curve* is bell-shaped (statistically measured by something called *kurtosis*) and symmetrical (statistically measured by *skewness*); that is, if you cut the curve in half, the right and left sides have the same shape. Kurtosis indicates how peaked or flat the bell-shaped curve is. The normally skewed curve in Figure 6.5 would be called "mesokurtic" in shape and would have a statistical value of zero kurtosis and zero skewness.

(a)

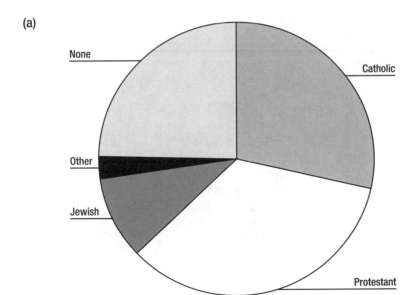

(b)

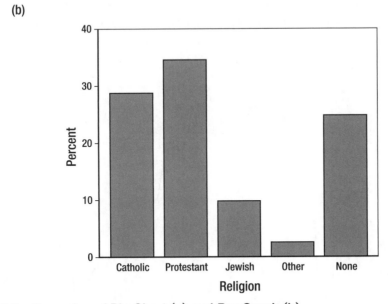

Figure 6.2   Examples of Pie Chart (a) and Bar Graph (b)

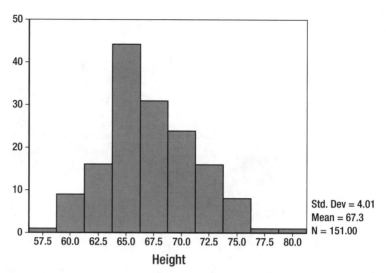

Figure 6.3   Example of a Histogram

If there are more scores bunched together on the left side or "negative" tail of the distribution and fewer on the right or "positive" side, then the curve is said to be *positively skewed.* A few high scores skew, or distort, the sample in favor of the positive end of the distribution. A *negative skew* is when few of the scores are at the left

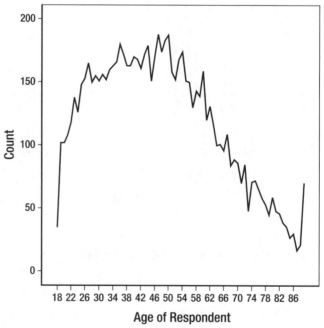

Figure 6.4   Example of a Frequency Polygon

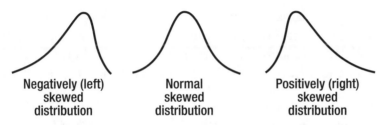

Negatively (left) skewed distribution　　Normal skewed distribution　　Positively (right) skewed distribution

Figure 6.5　Examples of Skewed and Normal Distributions

side or negative tail of the distribution and most are bunched at the higher end. A few extremely low scores distort the results in favor of the negative side of the distribution. This seems to go against intuition or the words we use when we talk about a "skewed" sample in everyday language, but think of a perfect bell-shaped curve as a piece of string, and whichever end you pull and stretch out to distort the curve a little tells you whether it's negatively or positively skewed. If you tug at the right side to include a few extreme high scores, you are skewing the sample in the positive direction.

Univariate Statistics

Generating statistical information about each variable in a study is another way to find out what you have and to understand more about the distribution of the variables in a sample. Of most importance is a *measure of central tendency,* which provides a quick summary of where the responses are clustered. Depending on whether the variable is nominal, ordinal, or interval/ratio, a mode, median, or mean is used. These measures can also tell you something about the distribution of a variable's values. When all three central tendency measures are equal, there is a perfect normal curve; all are in the center of the distribution. When the mean is higher than the median, there is a positive skew, because a few high scores distort the mean away from the median; a negative skew is indicated by a mean lower than the median, since a few low scores pull the mean down. Let's look at each of these measures.

The Mode.　For nominal data, the *mode* is used as a measure of central tendency. It is obtained by finding the most frequently selected value for a variable. A simple look at a frequency table, bar graph, or pie chart to see which value has the largest raw frequency or percentage of occurrence is all it takes to "calculate" this measure. For example, the modal political party in Table 6.1 is Democrat because it has the largest frequency count and valid percent. In Figure 6.2, you can also see that Protestant is the mode for the variable religion, even though it is not the religion of the majority of the respondents. Do not confuse the mode with the "majority" answer, which is a response that is more than 50 percent. The most frequently occurring value could have been selected by fewer than 50 percent of the respondents and still qualify as the modal response, but a majority response is always the mode. When there are two values that are selected with equal frequency, the result is a *bimodal* distribution.

The Median. If the values for the variable are ranked or ordered categories (ordinal data), a *median* is the ideal measure of central tendency to report in addition to the mode. The median—like the median that runs down the middle of a highway—is the halfway point, the value above which half the values fall, and below which the other half fall. It has virtually nothing to do with the actual values, just the number of values.

For example, imagine you asked five people to state their birth orders; you first list their responses in order: 1, 1, 2, 3, 3. The halfway point is the third response because this would place two below and two above that response. For these findings, the median birth order is 2. And if the fifth person instead said she was number 7 in her family of 8, the distribution would look like this: 1, 1, 2, 3, 7. Notice what happens: The median birth order remains 2. It is not affected by the size of the value, as the mean is. This is why it is a better statistic to use when you have a skewed sample of values that include extremely low or high scores like income. Not everyone would appreciate being told that the average income of people using Facebook includes Mark Zuckerberg's yearly take!

When you have an even number of responses—for example, only four people are surveyed and you found their birth order to be 1, 2, 3, 7—then you take the halfway point between the two middle values, in this case, between 2 and 3, which results in a median birth order of 2.5. How to calculate the median when the data are grouped—that is, when the ordered categories represent ranges (such as when value 1 represents ages 10 to 20, 2 equals 21 to 30, and so on)—is discussed in more detail in statistics books and available with an online search of the Internet.

Percentiles. The median is also called the fiftieth percentile. A *percentile* tells you the percentage of responses that fall above and below a particular point. So when you receive test scores on some national exam, like the SAT, and the results are at the eightieth percentile, it means that you scored higher than 80 percent of those taking it and lower than 20 percent.

Percentiles can be used to show the dispersion of scores for ordinal or interval/ratio data by finding the value at the twenty-fifth percentile (called the first quartile) and subtracting it from the value at the seventy-fifth percentile (the third quartile), resulting in the *interquartile range*. Percentiles can be broken down into any

BOX 6.1
BASIC DESCRIPTIVE MEASURES

Here is some output from SPSS using data from General Social Survey interviews.

Table 6.2   Descriptive Statistics

STATISTICS

Age of Respondent

N	Valid	1,401
	Missing	3
Mean		45.56
Median		42.00
Mode		34
Standard deviation		16.914
Variance		286.083
Range		71
Minimum		18
Maximum		89
Percentiles	20	30.00
	25	33.00
	50	42.00
	75	56.00
	80	61.00

Only 3 people out of 1,404 declined to reveal their age in the interviews. For those 1,401 valid responses, the following descriptive statistics were calculated. Because age is an interval/ratio measure, the mean and standard deviation are used. These data tell us much about one characteristic of the sample: 45.56 is the average age, 42 is the age half of the respondents are above and half are below (the median and fiftieth percentile), and the most frequently occurring age is 34. However, a frequency table needs to be viewed in order to find out exactly what percentage of the total are 34 (it's 3.1 percent of the 1,401 respondents).

The data also show that the youngest person in the survey is 18 and the oldest is 89, for a range of 71 years. To calculate the interquartile range, simply subtract the age or value at the twenty-fifth percentile from the value at the seventy-fifth percentile to get 23 years. Other percentiles can be used, for example, to show that 20 percent of the sample is over 61 or 80 percent are under 61 (the value at the eightieth percentile), whereas 20 percent are under the age of 30. The standard deviation is 16.9 years; note that this is the square root of the variance, 286.08. These numbers are meaningful primarily in comparison to another sample where age is used. If another survey reported that its mean age is also 45.56, but with a standard

BOX 6.1 CONTINUED

deviation of 20.3, then you would conclude that although it has the same average, its respondents are more widely dispersed around that mean. The ranges indicate this wider dispersion as well.

With these data, you could tell if the distribution of ages formed a normal curve by looking at the mean, median, and mode. If all are approximately the same number, then you have a normal distribution. Although the mean and median are within four years here, note that the mode is much lower. Because the mean is higher than the median (suggesting that there are some very elderly people pulling the mean higher from the median), and given that there is a clustering of respondents around a younger age (as the mode tells us), then it's likely there is a positive skew rather than a normal distribution of age. A positive skew occurs when most values cluster at the low end and a few values at the high or positive end pull the distribution in that direction. A histogram (shown in Figure 6.6), appropriate for continuous interval/ratio measures, illustrates the skew.

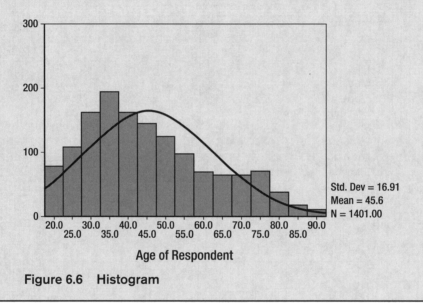

Figure 6.6   Histogram

number of categories, such as deciles to get the scores at every tenth percentile. A *range* indicates the spread of scores by subtracting the highest and lowest values. However, the range is affected by extreme scores unlike interquartile ranges, which are better suited when there are some very high and very low values in the distribution. Many college guidebooks present comparisons of SAT scores using the interquartile range.

The Mean. The most sophisticated measure of central tendency, and one that forms the basis of advanced statistics, is the arithmetic mean. You have calculated this many times, and it follows you around school as the grade point average (GPA).

The mean is the sum of the values divided by the number of values and is most suitable for interval/ratio variables. $\bar{X}$ (pronounced X-bar) is used for samples, and mu (μ) is used for means when you have surveyed every element in the population.

$$\bar{X} = \frac{\sum X}{N}$$

Calculating the mean for some ordinal scales (such as Likert ones) that look like equal-appearing interval scales is acceptable. When the interval/ratio measure is

---

BOX 6.2
DESCRIPTIVE STATISTICS IN A PUBLISHED REPORT

The Pew Research Center (2011b) surveyed 2,277 Internet users ages 18 and older who use text messaging on their cell phones. They were asked about the number of text messages they sent and received per day. Here are some findings from this study:

NUMBER OF TEXT MESSAGES PER DAY		
	MEAN	MEDIAN
All text messaging users	41.5	10
Gender		
Men	40.9	10
Women	42.0	15
Age		
18 to 29	87.7	40
30 to 49	27.0	10
50 to 64	11.4	3
65+	4.7	2
Education level		
Less than high school	69.4	20
High school diploma	45.4	15
Some college	53.0	15
College+	23.8	10

$N = 2,277$

Remember, the mean is a mathematical calculation affected by extreme scores, while the median tells us simply what number is at the fiftieth percentile, that is, where half the respondents are above and below that point. So, the univariate statistics tell us that half these respondents send or receive fewer than 10 texts a

BOX 6.2 CONTINUED

day, and half send or receive more than 10 a day. Yet, these same people average 41.5 messages a day. This suggests that there are several people sending/receiving quite a large number of texts, which skews the results in a positive (higher) direction.

We can also see by these data how other differences begin to emerge when we look at bivariate results (discussed in more detail in Chapter 7). For example, men and women seem to send/receive texts about the same amount on average (men send/receive almost 41 a day; women have a mean of 42 a day), but half the women respondents send/receive more than 15 texts a day, compared to half the men with more than 10 a day, as the median indicates.

Notice the big difference by age: Americans age 65 and older average fewer than 5 texts a day (with a mean of 4.7 to be exact), and only half send/receive more than two texts a day. Compare this to the 18- to 29-year-olds who average almost 88 texts a day, with half of these respondents sending/receiving more than 40 messages a day.

Using the median along with the mean gives you more information than either does alone. Consider another bivariate finding (educational level and number of texts are the two variables): Those with a high school diploma and those with some college have the same median number of texts. Half of both subgroups of respondents send/receive fewer than 15 messages a day (according to the median); yet those with some college average 53, almost 8 more per day than those respondents with only a high school diploma. This tells us that the distribution of text messages among the "some college" respondents is more positively skewed than the distribution of those with a high school diploma; that is, several more people in the "some college" group are texting a lot more and thus pulling the average higher than the "high school diploma" group.

What is missing from these data that would give us even more information about the distributions for each subgroup? It would help to have, along with the means, the standard deviations, a very useful statistic for interval/ratio data.

discrete rather than continuous, the mean often produces a peculiar number, such as the 2.3 mean number of children in families, or the 38.5 average number of books borrowed at the library per day (can you really borrow just half a book?).

Standard Deviation. Besides describing the dispersion of values using a range, such as the interquartile range, a more powerful measure of dispersion, the *standard deviation,* is available for interval/ratio data. Think of it as the average variation of all the values from the mean. With the standard deviation, we can compare similar variables in different samples or in the same sample at different points in time. The larger it is from zero, the more dispersed the scores are in the sample for that particular variable.

The standard deviation number itself is in the units of the values of the variable—for example, scores on a 100-point reading test—and cannot be compared to a standard deviation calculated on a different variable, like inches for height. They are most useful when comparing similar measures between two (or more) different groups, or two (or more) sets of similar measures for the same group. For a quick sense of whether a particular interval/ratio variable in your study is indeed a variable, this statistic provides that information. The further away the number is from no deviation of zero, the more dispersed the scores are for that variable in your sample. Box 6.3 illustrates how the standard deviation is used and calculated.

BOX 6.3
CALCULATING THE STANDARD DEVIATION

The standard deviation (s for sample statistics, σ [sigma] for population parameters) is calculated by taking each score (X) and subtracting (deviating) it from the mean. Because the mean is the perfect mathematical middle of a set of values, when you deviate the scores from the mean you notice that, when added together, these deviations always equal zero. If one score is -2 below the mean, then another one is $+2$ above the mean, and so on for every value in the distribution. This creates problems because to calculate a mean you have to add up the scores and divide by the number of scores. If the sum of those scores equals zero, good luck in doing any division!

$$s = \sqrt{\mathrm{var}} = \sqrt{\frac{\sum (X - \overline{X})^2}{N - 1}}$$

So, the calculation for the standard deviation requires that the deviations be squared to eliminate the negative numbers; then they can be added and divided by the number of scores minus one (sometimes called *df*, or the *degrees of freedom,* a concept explained in Chapter 7). The resulting number is called the *variance* (s^2), a core concept of statistical analysis. A good deal of the time, we want to understand why data vary in our sample. We usually want to explain the variance in our dependent variables in terms of the variance in our independent variables. We have to remember, though, that we sort of arbitrarily squared all those differences in order to calculate the variance, so just to "undo" what we did, we take the square root of the variance and the resulting number is called the standard deviation.

Let's take a teacher who found that the 30 students in her third-grade class had an average reading score of 75 and another teacher discovered that his 30 third graders also had an average reading score of 75. These findings wouldn't tell you much more than that both classes seemed to be at the same level of reading ability. Yet, every student in the first class might have scored exactly 75 to achieve an average of 75 (add up 30 scores of 75 and divide by 30), while in the second class 15 of the students may have scored 60 and another 15 may have scored 90, also resulting in an average of 75. The second class has a much larger dispersion of scores: The students range from 60 to 90. Comparing the two means by themselves wouldn't uncover this very unusual and important difference.

What we need then is the standard deviation to give us this information. In the first class, the standard deviation is zero. This is obtained by subtracting all 30 scores of 75 from the mean of 75, resulting in lots of

BOX 6.3 CONTINUED

zeros, and when zero is squared, you still get zero. The sum of zero divided by 30 remains zero even after you take the square root of zero. See what happens when you use zeros in multiplication or division!

On the other hand, in the second class, when you subtract 15 scores of 60 from 75, you get many scores of –15, and when you subtract the other 15 scores of 90 from 75 you see many scores of +15. Square all those numbers and you now have 30 values of 225 to add together. Divide that total by 29 ($N-1$) and take the square root of the new number to get the standard deviation of 15.26. This is a much larger number than zero.

When comparing the two classes' reading scores, the standard deviation tells us that the students in the second class are more widely dispersed in their reading ability than the students in the first class. For the first class, reading scores are a constant, not a variable, as a frequency table or histogram would have also illustrated.

THE NORMAL CURVE AND Z-SCORES

Standard deviations are also important for figuring out where exactly a particular value or a sample of scores is relative to the mean. If the distribution of scores form a normal curve—indicated by the frequency curve or by looking to see if the mean, median, and mode are the same number, as they are when the distribution is normal—you can compare any score with others by standardizing them using *z-scores*. These are used to locate a particular value in a distribution of values, to get its percentile rank, or to compare it with a score measured in different units. Let's say you want to compare respondents' grades on a classroom reading test with their verbal aptitude scores on a national test. Because they are measured in different units, you must first translate the values to z-scores. Box 6.4 illustrates how this works.

Statistical Significance, Confidence Intervals, and the Central Limit Theorem

Z-scores are used for calculating the percentiles of individual scores. Therapists, guidance counselors, and others working with data about individuals can see if a client, for example, is significantly different from the norm on a psychological test. You could also see how you stand in comparison with others who took the same test, especially if your teacher is "grading on the curve," that is, using a normal curve to determine grades.

Besides giving us a percentile, z-scores assist in determining confidence intervals for understanding probability levels of significance and in determining sampling error. A z-score tells us the probability of obtaining a score by chance. If 50 percent of the scores are above the mean in a normal distribution, then the probability

BOX 6.4
CALCULATING Z-SCORES

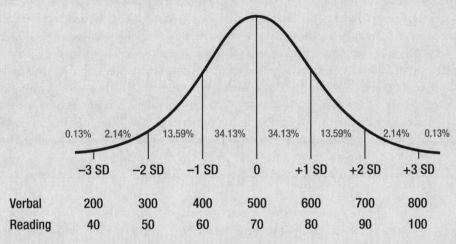

	−3 SD	−2 SD	−1 SD	0	+1 SD	+2 SD	+3 SD
Verbal	200	300	400	500	600	700	800
Reading	40	50	60	70	80	90	100

Figure 6.7   Normal Curve Distribution

Imagine that a student scored 600 on a national SAT verbal aptitude test that goes from 200 to 800 points. Let's assume the mean for that test is 500 and the standard deviation is 100. Now let's also imagine that a student has a score of 80 on a classroom reading test where 70 is the mean and the standard deviation is 10. Assume that both distributions form a normal curve (Figure 6.7). Is the student consistent in how she stands within each of these normal distributions? Is a score of 600 equivalent to a score of 80? To answer that question, z-scores are calculated. A z-score is obtained by subtracting the mean of the distribution from a score and dividing the result by the standard deviation, in order to "standardize" it.

$$z = \frac{x - \overline{x}}{s}$$

For the verbal aptitude score, 600 − 500 = 100, then 100/100 = 1. This student has a z-score of 1 for a test score of 600. For the reading test, 80 − 70 = 10, then 10/10 = 1. We now know that a verbal score of 600 and a grade of 80 on the reading test are comparable standardized scores, even though they are in different units of measurement. Both are one standard deviation unit above the mean. If it were a minus result, it would tell us the score is below the mean.

A z-score also gives us the *percentile* for the score. Based on the mathematics in the formula for the normal curve (available in advanced statistics books and on the Internet), a normal distribution of values has a mean of zero and a standard deviation of one. Approximately 99.7 percent of all the scores in a distribution (the area under the normal curve contains all the scores) fall within three standard deviations above and three standard deviations below the mean; approximately 68 percent of all scores fall within plus or minus

BOX 6.4 CONTINUED

one standard deviation of the mean; and 95 percent fall within two standard deviation units of the mean. Using this information and a table of z-scores and percentages (also found in most statistics books and on the Internet), we can figure out the exact percentile of each score in the distribution.

For example, 34.13 percent of all the scores fall between the mean and one standard deviation above or below the mean. Since the curve is symmetrical, we also know that 50 percent are below the mean; therefore, a z-score of 1.0 tells us that 84.13 percent of all the scores are below that one (50 + 34.13). In other words, a verbal aptitude score of 600 has a percentile rank of 84.13. Conversely, 15.87 percent of all the scores are higher than 600. We can also say that approximately 68 percent of all the scores fall between 400 and 600, since 34 percent fall between the mean (in this case, 500) and one standard deviation unit (in this case, 100 points) below the mean and another 34 percent fall within one unit above the mean.

When a score is exactly the same as the mean of the distribution—say, a score of 500 on the verbal test—then the z-score would be zero. In this case, 50 percent of the scores are above and another 50 percent are below 500. You can see that the mean is also the median in the normal curve.

of finding someone with a verbal aptitude score above 500 is 50 percent or, conversely, for finding someone below 500. If the percentile is 84 percent, then the odds of finding someone with a score above 600 is 16 percent, 84 percent for finding someone below 600, and so on.

Statistical Significance. One of the conventions in social science research is to declare that a finding is *statistically significant* if the probability of obtaining a statistic by chance alone is less than 5 percent. This can be either 5 percent at one end, or tail, of the normal distribution for one-directional tests, or 2.5 percent at the low or negative end *plus* 2.5 percent at the high or positive tail of the distribution for two-directional tests. A *two-tailed* or *two-directional test of significance* exists when we allow two chances in our hypothesis for the outcome to be different: either differently high or differently low from what is expected. For a score to be significant, then, it must be lower or higher than the z-score at which 2.5 percent of the scores fall above or below. The top 2.5 percent of values in a normal distribution have a z-score higher than +1.96 and 2.5 percent would be lower than –1.96, or above a z-score of approximately 2 and below a z-score of –2.

If we set our significance level (often called the *alpha level*) to the probability of obtaining a statistic by chance 1 percent of the time, that is, in the top .5 percent or the bottom .5 percent for two-directional hypotheses (.5 + .5 = 1), then we would use a 2.58 z-score (plus or minus) as the point above or below which significance would occur. Consider a distribution of verbal scores where 500 is the mean and 100 is the standard deviation: A verbal score of 758 or higher would be in the top .5 percent. Each z-score unit in this distribution is "worth" 100 standard deviation points, so 2.58

times 100 results in 258 points above the mean of 500. A score of 242 or lower would be in the bottom .05 percent (500 minus 258) of the distribution of all verbal scores. The probability of selecting someone randomly from a normally distributed population whose verbal score is above 758 or below 242 would be less than 1 percent; selecting someone with those scores could then be called a statistically significant outcome. Notice that we get two chances to be significantly different from the norm: either very high scores or very low ones. That's what is meant by a two-tailed test of significance.

Alpha significance levels are usually represented as $p < .05$ or $p < .01$ or $p < .001$, meaning the probability of obtaining that statistic (or score) by chance is less than 5 in 100 (5 percent), 1 in 100 (1 percent), or 1 in 1,000 (.1 percent), respectively. A single asterisk (*) typically appears in tables of data to signify that the .05 significance level has been reached; two asterisks are used for .01, and three for the .001 level.

If we want to see whether a particular score is significantly *higher* than the average—that is, it would be significantly different only if the score were higher but not lower than the average—then we would apply a *one-tailed* or *one-directional test* of significance. In this case, to be statistically significant, the score should be higher than 95 percent of the values and in the top 5 percent with a z-score of 1.64, or –1.64 if looking for scores significantly lower than the average and in the bottom 5 percent. In short, when the probability of obtaining a particular statistic by chance is less than the cutoff point established, we can state that the hypothesis or research question being tested is statistically significant. Call the media to announce your significant finding and tweet your excitement to everyone!

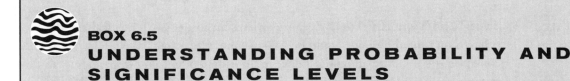

BOX 6.5
UNDERSTANDING PROBABILITY AND SIGNIFICANCE LEVELS

When calculating probability and using "or" as the criterion, you *add* the odds. When you use "and" as the criterion, you *multiply* the odds. For example, if you were asked to pick a red card from a deck of cards (26 chances out of 52 or 26 ÷ 52 = .50) *or* a black card (also .50), then the probability of being successful (significant) is .50 + .50 = 1.00. In short, you have a 100 percent chance of picking a red *or* black card. This should come as no surprise!

But if you asked what the odds are of picking a black card (.50) *and* an ace (4 chances out of 52 = .077), then the probability of picking a black ace is .077 × .50 = .038. There is a 3.8 percent chance you will pick a black ace from a deck of cards. Another way of figuring this is to calculate the odds of getting the ace of spades *or* the ace of clubs using the additive rule. The odds of picking the ace of spades is 1 out of 52, or .019. The odds of picking the ace of clubs is also .019. Therefore the odds of picking either the ace of

BOX 6.5 CONTINUED

spades *or* the ace of clubs is .019 + .019 = .038. In short, you have two chances out of 52 to pick a black ace, or 2 ÷ 52 = .038.

The key point to remember here is that when we are testing to see what the odds are of obtaining a statistic or value by chance that is significantly higher *or* lower than the mean at the .05 alpha level (for a two-tailed hypothesis), then we are looking for a value that is either in the top .025 area of the normal curve (the top 2.5 percent) *or* in the lower .025 area in order to get an added probability of .05, that is, .025 + .025 = .05. When we do, we can declare that the statistic or value is a significant finding, just as we would in randomly pulling a black ace from a deck of cards, because the odds of doing so are certainly less than .05. As we just saw, the probability is .038 and therefore $p < .05$.

Here's another easy way of remembering significance levels: Think about 1,000 people standing around flipping coins. Let's say you ask them to flip a coin 10 times. At what point do you accuse some of them of using a phony coin? If they get five heads or tails? Nah, that's a pretty likely outcome. When they get three or four heads or tails, or six or seven or what? Those seem fairly reasonable outcomes as well. However, the odds of getting 10 heads or 10 tails by chance when you flip a coin 10 times is .001. That is, only one person in that group of 1,000 is likely to do so by chance alone. Any more and you might be suspicious that some people have magic coins.

The odds of getting nine heads or tails are around .01, or 1 in every 100 people flipping a coin 10 times. You wouldn't expect many more than 10 people in your sample of 1,000 to get nine heads or tails. And the chance of flipping eight heads or tails is approximately .044, or less than .05 using our criterion of significance.

In other words, you would say that it would be a statistically significant event if someone got 8 ($p < .05$), 9 ($p < .01$), or 10 ($p < .001$) heads flipping a coin 10 times since the odds are so low that these could occur by accident. So when you see significance levels, translate them into coin flips and ask yourself if the statistical findings are as rare as these coin flip results. If they are, you have a statistically significant finding.

Type I and Type II Errors. Because a decision to declare statistical significance is based on probability, we could be wrong some of the time. We might incorrectly conclude there is a relationship between an independent and dependent variable in the population from which the sample was chosen when there really isn't one. Finding a relationship when there isn't one could occur the same percentage of time established by the alpha level of significance. For example, if we wanted to declare statistical significance at the .05 level, then the probability of the statistic occurring by chance is less than 5 percent or, put another way, we are 95 percent confident it is not due to chance but instead to a real finding.

However, this could be one of those chance times that a statistic of the magnitude calculated happened accidentally and was not due to actual findings and relationships in the population. The likelihood, then, of declaring a relationship statistically significant when it is not is the value of alpha; this is called a *Type I error*. In the words of the null hypothesis, a Type I error is rejecting a null hypothesis that is truly null, when we should have accepted it. There is no relationship between the independent and dependent variables, but we rejected that by mistake and declared there is a relationship. Normally, when a statistic's significance level is less than .05, or whatever other number was selected, we say that we have a statistically significant finding and we reject the null hypothesis. That is, we reject that there is no relationship and declare the alternative hypothesis that there is a relationship. In this case, we should have accepted the null hypothesis of no relationship.

We hold a press conference to announce our findings, yet we should be aware that our outcome could be the result of chance and not due to the independent variable, as the statistical calculations have led us to believe. By setting a probability strict enough to limit the number of Type I errors, we become more reassured that our results are genuine and not likely to be due to chance alone. Thus, many researchers choose .01 or .001 as stricter levels of significance that are more difficult to achieve. The odds of making a Type I error and declaring a relationship when there really isn't any, in these cases, is less than 1 in 100 (1 percent) or 1 in 1,000 (.1 percent).

On the other hand, we might be making a *Type II error* instead and accept a null hypothesis that should be rejected, that is, declaring that there is no relationship between our variables when in fact there really is a statistically significant one. These two errors are intertwined: Decreasing a Type I error increases the possibility of making a Type II error. The goal of testing hypotheses is to minimize errors when making inferences about a population from a sample, but convention encourages us to use at least .05 as the level of significance ($p < .05$) when testing statistics, depending on what is being tested and for what reasons. Exploratory research may allow a lower standard, while predictive research looking at the outcomes of a new kind of life-saving drug or public policy program, for example, might require a more stringent test, such as $p < .001$.

The key point to remember is that any results and statistics never prove a hypothesis or research question. Statistics suggest a finding that is tentative; we just never know for sure whether what we found occurred by chance or is due to actual influence of the independent variable on the dependent variable. All we can say is that the probability of a relationship happening by accident is less than a certain standard we set.

Central Limit Theorem. Making inferences about a population and establishing probabilities assume a normal distribution. What happens, though, when the distribution of a variable is not normal, as is usually the case? There are other distribution shapes and appropriate statistics that can be used for other models, but these are beyond the scope of this book. Luckily, though, the distribution of statistics (like means) is assumed to be normally distributed. If you were to plot on a graph the reading test means of all possible samples of people taken from a population, the distribution would approach a normal curve, especially the larger the sample size is. This is something in reality you wouldn't do, because if you had the time and money to gather all possible samples of a certain size, you would end up with a survey of the entire population. If the population were small enough for you to survey, then you wouldn't have a need to sample in the first place.

What we are talking about is a theoretical idea known as the *central limit theorem* and something useful in understanding sample size, sampling error, and confidence limits. It states that a distribution of sample means—again, not to be confused with the distribution of individual scores from one sample—will approach a normal curve, the larger the sample size and the larger the number of samples taken. And the mean and standard deviation (here called the *standard error of the mean*) of all the sample means will be the true population mean (μ) and population standard deviation (σ).

Consider a situation in which we happen to know what the population parameters are for a group of people. Imagine that the entire population of some small village is 1,495 people and has an average age of $\bar{X} = 46.23$ and a standard deviation of $s = 17.42$. As illustrated previously, the distribution of age is not normally distributed; it has a slight positive skew. Now, let's take random samples from that population of size five (see Table 6.3). If we were able to take all possible combinations of samples of size five, calculate the mean age for each of those samples, plot them on a graph, and then calculate a mean of all those means, the graph would begin to look like a normal curve and the mean of all those means would be equal to 46.23, the true population mean.

Because the sample size is so small (only 5 out of 1,495), many of the samples are way off from the true mean of 46.23. Yet, even without showing all possible samples of size five from the population, a mean of 45.96 for just five of these sample means gets pretty close to the true population parameter. What if the one random sample of five we actually generated turned out to be sample three? The odds are great that, with small sample sizes, the sample statistics (in this case the mean and standard

Table 6.3 Samples of 5

	SAMPLE SIZE (N)	MEAN ($\bar{X}$)	STANDARD DEVIATION (s)
Sample 1	5	44.8	19.15
Sample 2	5	35.2	17.21
Sample 3	5	56.0	16.51
Sample 4	5	40.0	10.07
Sample 5	5	53.8	15.51

deviation) will be further away from the true population parameters and we are more likely to end up with a sample that does not represent the population. There would be greater sampling error, hence the justification to gather samples of larger sizes (and good-quality ones using random probability sampling methods).

The central limit theorem argues that larger sample sizes will result in a distribution of sample statistics that is closer to a normal curve with most of the values close to the mean (a curve that is narrow and peaked), and therefore we would be more confident in the accuracy of predicting the population information and make fewer errors doing so.

Now consider random sample sizes of 50 from the same population in Table 6.4. Here, the mean of the five means is 46.44, almost exactly the true population mean of 46.23. And that is with only five samples. In other words, any one of these samples would be a much better estimate of the true population mean than any one of the samples that had only five people in it. Still not convinced? Then take a look at random sample sizes of 100 in Table 6.5.

The mean of the sample means is 46.76, pretty close again to the actual population mean of 46.23. Also note how close the standard deviations are to the true population standard deviation of 17.42. Any one of these samples would provide a more accurate estimate of the population parameters than any of the samples with sizes of five. Remember, in actual research we only survey one sample. Notice that the increase in accuracy from sample sizes of 50 to 100 may not be worth the extra time and money; doubling the sample size did not increase the number of samples that

Table 6.4   Samples of 50

	SAMPLE SIZE (N)	MEAN ($\bar{X}$)	STANDARD DEVIATION (s)
Sample 1	50	46.0	16.37
Sample 2	50	46.8	16.79
Sample 3	50	46.5	16.38
Sample 4	50	44.6	13.80
Sample 5	50	48.3	17.16

Table 6.5   Samples of 100

	SAMPLE SIZE (N)	MEAN ($\bar{X}$)	STANDARD DEVIATION (s)
Sample 1	100	46.4	17.14
Sample 2	100	46.5	17.59
Sample 3	100	47.2	17.70
Sample 4	100	44.6	16.05
Sample 5	100	49.1	17.67

were closer to the true population. In general, as we increase the random sample size, more of the samples' means tend to be closer to one another and the true mean, resulting in fewer sampling errors.

In Chapter 5, we discussed how large a sample should be for a good study. One of the points made was that the larger the random sample size, the more likely it is to capture the diversity that exists in the population. This idea is based primarily on the concepts just described about the central limit theorem. Similar calculations are made by professional research organizations when determining sample size. For example, one formula used to determine sample size takes sampling error into account: margin of error equals 1 divided by the square root of the sample size. Choosing a sample size depends on the population size, its heterogeneity on a variety of key characteristics being studied, the amount of sampling error you're willing to tolerate, and the amount of money and time available to do the research.

Confidence Intervals. The central limit theorem is a theoretical concept because in reality no one takes all possible samples from a population. However, it does help in determining whether the one sample we actually have is representative of the population, that is, whether the sample mean and standard deviation are fairly accurate estimates of the true population mean and standard deviation. Using *inferential statistics,* we infer (arrive at some conclusion about) the population parameters from the sample statistics with some degree of confidence. We can do this since the normal curve tells us with z-scores and percentiles how far off the statistics are from the mean. Not unlike political polls that present results in terms of plus or minus four or five percentage points of the true population percentage, what we also do is determine how confident we are that the true mean of the population (or other statistic) is within a range of plus or minus a certain number of points.

This range is referred to as the *confidence interval,* and the numbers at the beginning and end of the interval are called the *confidence limits.* You see this range in the output results performed by most computer statistical programs. If we use the traditional cutoff point for significance of .05, we can state that we are 95 percent confident that the true population mean—or whichever other statistic from the sample we are using to infer the population parameter—is within this confidence interval range. It states that if we were to get 100 samples of the same size as the one we just did and calculate the confidence limits for each one of them, 95 of those confidence limits would contain the true actual population mean. We could also set the alpha level at .01, or 1 percent, and use a z-score of plus or minus 2.58 to calculate the confidence interval if we want to be 99 percent confident that the true population mean is within that range. In any case, we will never know the true population parameter unless we do a survey of every element in the population, so our results are always an estimate and subject to some sampling error.

What the central limit theorem provides us with, then, is proof that even when an individual variable is not normally distributed in a sample, if large-enough sample sizes are taken, the distribution of statistics (like the mean) from those samples will be normal. This fact allows us to use probabilities in making conclusions and inferences about the population from which the one sample we actually have comes. Our inferences have fewer errors, the more confident we are that the true number is within a particular range. And we make many more errors, the smaller the sample size is, because the distribution of statistics from small random samples has a standard deviation (standard error) that is larger than the true population's dispersion.

The central limit theorem tells us that larger sample sizes produce distributions that have more limited dispersions, ones in which the normal curve is narrow and peaked. We will make fewer errors in estimating the population parameters if we have a sample from a theoretical distribution of all samples that is normally distributed this way since we know that the odds of obtaining a sample that is statistically very different from the norm (the mean, for example) are 5 in 100, 1 in 100, 1 in 1,000, or whatever alpha level we set as our standard of statistical significance.

Considerations in Descriptive Analysis

Practically speaking, what we need to be aware of when reviewing the output from various statistical procedures is the information about probability or significance levels, confidence intervals, standard error, and sample size, as described in the next three chapters. These numbers are our guide to making more accurate inferences or generalizations about the characteristics of the population from which we drew our random sample. Without a random sample, however, generalizations about a population are not possible, and inferential statistics cannot be calculated. Yet, we can still use many of these statistics and graphs to describe what we have in our study.

However, it's critically important to remember that research ethics play a central role at this stage of the research process. Making conclusions about a population using data collected from a convenience and nonrepresentative sample demonstrates not only a lack of understanding about sampling but also unprofessional ethical choices. A researcher must always report the findings accurately, use the appropriate statistics, and present them correctly in graphs and tables. Numbers can easily be manipulated by using exaggerated scales on a graph (beginning the y-axis units at 50 and increasing with units of 5, for example, as opposed to starting at 0 and increasing with units of 10), by selecting a mean when a median should be used, or by presenting the results as if they applied to an entire population rather than just the sample. The ethical use of statistics and the accurate interpretation of them must be kept in mind as you learn in the next three chapters how to do more complex data analysis.

Displaying and describing the basic statistics for each of our variables is the first step in analyzing data. These univariate statistics help us determine if the items from our questionnaire are actually variables and useful for later data analysis. They also allow us to describe the characteristics of the sample through displays of the demographic findings. Once we determine which items are variable enough for further analysis, we can begin to evaluate the research questions and hypotheses we developed using concepts of inferential statistics and probability levels. To do this, we need to explore relationships between two or more variables at a time. The next chapter discusses the many ways to assess whether variables are correlated.

REVIEW: WHAT DO THESE KEY TERMS MEAN?

Alpha levels	Kurtosis	Positive and negative
Bar graphs	Mean	skews
Central limit theorem	Measures of central	Range
Confidence limits and	tendency	Standard deviation
interval	Median	Standard error of the
Frequency curves	Mode	mean
Frequency tables	Normal curve	Statistical significance
Graphs and charts	One- and two-tailed	Type I and Type II errors
Histograms	tests	Univariate analysis
Inferential statistics	Percentiles	Values and variables
Interquartile range	Pie charts	Z-scores

TEST YOURSELF

1. For each of the following variables in a study, list the one best measure of central tendency you can use to describe the data and a graph to depict the distribution of values (pick any one if more than one can be used).

	Levels of Measurement	Graph/Chart
a. SAT scores		
b. Race/ethnicity		
c. Skewed number of hours studied in the past week		
d. Type of car owned		

2. Here are some results using data from 151 students who reported their height (in inches):

Mean: 67.3   Median: 67   Mode: 68   Standard deviation: 4

a. How can you tell if this is approximately a normal curve?

b. Using the standard deviation, calculate the following:

   i. What percentage of the sample is between 63.3 and 71.3 inches?

   ii. What height would put you in the shortest 16 percent of the class?

   iii. What percentage of the respondents are taller than 67 inches?

   iv. If you set the significance level at 5 percent ($p < .05$), and you were to randomly select one respondent from the sample to find someone "significantly different from the mean," approximately what height would it take for this respondent to qualify as statistically taller or statistically shorter than the mean? Is this a two-tailed or one-tailed research question?

   v. Assume this is a random sample from a population: Between approximately what heights would you be 95 percent confident that the true population mean is?

INTERPRET: WHAT DO THESE REAL EXAMPLES TELL US?

1. In a published academic article, McMahon et al. (2009: 272) presented fourth- and fifth-grade low-income students' perceptions of the classroom and school environment and the relationship of those views to academic outcomes. The researchers used a standardized measure called "My Class Inventory" composed of five subscales: satisfaction with the class, friction among students, competitiveness among students, difficulty of the work, and cohesiveness of the class. Because answers to each item were 0 = no and 1 = yes, scale scores ranged between 0 and 1. Here are the results:

	MEAN	STANDARD DEVIATION
Satisfaction	0.64	0.29
Cohesiveness	0.54	0.35
Friction	0.44	0.32
Competitiveness	0.55	0.31
Difficulty	0.20	0.22

a. What do the means tell us in terms of the five subscales? Are the elementary school students satisfied with their class and school environment? Do they find the work difficult? In short, how would you put into words what this table of means tells us?

    b. What do the standard deviations say? How diverse are the students' perceptions? On which measure is the range of responses more varied? In short, how would you put into words what this table of standard deviations tells us?

2. The National Health and Social Life Survey asked respondents to indicate how many sexual partners they have had in their lives since the age of 18. According to the findings (Laumann et al. 1994: 180), the minimum number reported by the 3,126 respondents was 0, and the maximum was 1,016. The researchers found the median number was 3 and the mode was 1.
    a. Why did they use the median for the level of central tendency and not the mean?
    b. Put into words what a median of 3 and a mode of 1 tell us.
    c. What is the range for these data? What else would you need to know to interpret these findings?

CONSULT: WHAT COULD BE DONE?

You've been hired to do the data analysis for a survey conducted at a local business. The questionnaire contains the following variables, and you need to advise the business which ones are good for further analysis in its study.

1. What do you do first? Describe the steps needed to be taken at the start of data analysis.
2. Which measures of central tendency and what kinds of graphs would be best to use to describe the following variables?
    a. Type of job performed in the organization
    b. Hours worked per week
    c. Seniority ranking (oldest employee to newest)
    d. Monthly salary
    e. A checklist of employee benefits (health plan, dental insurance, child care, etc.)

DECIDE: WHAT DO YOU DO NEXT?

For your study on how people develop and maintain friendships, as well as the differences and similarities among diverse people, respond to the following items:

1. For each variable in the questionnaire you developed in the previous chapter, describe which measures of central tendency, other statistical descriptions, and graphs would be best to use during the first phase of data analysis.
2. If you have actually collected data, now is the time to code and enter the data into a statistical program and begin data analysis with descriptive statistics.

ANALYZING DATA

Bivariate Relationships

7

> The invalid assumption that correlation implies cause is probably among
> the two or three most serious and common errors of human reasoning.
>
> *—Stephen Jay Gould, scientist*

LEARNING GOALS

Understanding bivariate statistical analysis is the focus of this chapter. Central to this is learning how to read and construct cross-tables of data and deciding which statistics to use to measure association and correlation. By the end of the chapter, you should understand how to reject or accept a hypothesis using the appropriate statistics to assess bivariate relationships. You should also be able to put together cross-tables and interpret them clearly in words.

Figure 7.1 Statistical Decision Steps

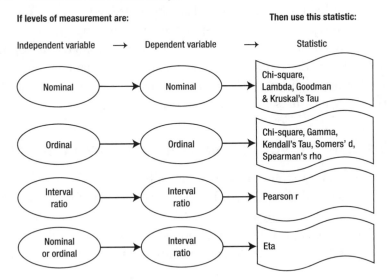

After you evaluate your univariate data analysis and feel confident that you have variables with good distributions, it is now time to begin investigating the bivariate relationships you proposed in your research questions and hypotheses. *Bivariate* data analysis assesses the association between two variables. If you are interested in studying three or more variables at a time, then multivariate analysis is required (see Chapter 9). Remember, if for some reason any of your variables approximate a constant—that is, almost everyone has selected only one or two of the values for that variable—then you must eliminate them from further analysis. For example, when 90 percent of the people who completed the questionnaire agree or strongly agree that they are satisfied with their current job, then job satisfaction is no longer a variable in your study unless the sample size is very large.

Keep in mind what the objective is: to demonstrate whether the independent variable is exerting significant influence on the dependent variable in order to infer something about a population from which your sample was selected. You want to see if the variation that exists among the respondents in the dependent or outcome variable can be explained or predicted by the variation that exists in the independent variable. For example, you might want to establish whether studying more leads to higher grades. When there is some statistically significant relationship, you have fulfilled one of the three requirements to establish cause and effect (see Chapter 1). Once you have eliminated alternative explanations and can show that the independent variable occurred prior in time to the dependent variable, then you hold a press conference and declare you have found a cause for the occurrence of the dependent variable.

Another objective may simply be to describe how the dependent variable varies according to different categories of the independent variable. You might be interested in showing whether political party affiliation differs between men and women or among Latinos, Asian/Pacific Island people, and African Americans. Or you would like to report how many respondents in your study are young, middle-aged, and elderly and who are from rural and urban areas, such as young urban, young rural, elderly urban, elderly rural, and so on. This chapter focuses on developing skills for accomplishing these objectives through bivariate data analysis and for understanding measures of correlation and association.

PRESENTING NOMINAL AND ORDINAL DATA IN TABLES

Using your research questions and hypotheses, begin by making a list of the variables in them. Next, label each variable in terms of its level of measurement (nominal, ordinal, or interval/ratio), and—based on what follows in this and the next two chapters—decide the appropriate way of presenting the data visually with tables or graphs. Finally, select the most relevant statistics to assess what you have found.

Similar to the way you show the distribution of a single variable, constructing frequency tables for both variables simultaneously is a good way to start analyzing the variables in your hypotheses. What is the occurrence of one variable in terms of the other one? Is one variable's distribution contingent on the other's variation? Is the dependent or outcome variable associated with the independent or predictor variable in any way? To answer these kinds of questions visually, a table crossing the two variables is made. These are usually referred to as *contingency tables* or *cross-tabulations* (*crosstabs,* for short). They illustrate the number and percentage of occurrences in the sample of each value of one variable simultaneously with each value of the other variable. Remember, variables are usually the questions in your survey, like "age"; values are the answers to the questions, like "under 25" or "between 40 and 50."

Crosstabs are ideally suited to nominal or ordinal measured variables or to interval/ratio data with a very limited number of discrete values. Each value (or category) of one variable constitutes the columns in the table, and each value (or category) of the other variable makes up the rows of the table. It is easy to remember which is which: Just as *columns* hold up a building, they hold up a table; just as you move across the *row* to get to your seat in the movie theater, you move across the rows in a table. These are similar to the concepts used in a spreadsheet program, like Excel or SPSS.

Ideally, the number of rows and columns in a table should be small enough to see all of them at the same time on a screen or sheet of paper, thereby restricting crosstabs to mostly nominal or ordinal measures. Interval/ratio data with a limited number of values can also be used, but imagine a table that has a row for every single age that exists in your sample of 18- to 89-year-olds! The size of a table is referred to in terms of its rows and columns, in that order, such as a 3 × 2 table (three rows and two columns). *Cells* are the locations where each row's values and each column's values intersect, and the number of them is quickly calculated by multiplying the number of rows by the number of columns: A 3 × 2 table has six cells, for example. A cell's location is also given in terms of its row and column number, so you can refer to the data in a cell as being "in row 3, column 3."

Another convention is that the percentages for each value (or category) of the independent variable should add up to 100 percent; if the values of the independent variable are columns, as is recommended, then each column should total 100 percent. In this way, you standardize the responses when the numbers of responses for each value of the variable are not equal. This allows you to make a fair comparison about the frequency of occurrence of each value of the independent variable in terms of each value of the dependent variable.

Let's say you are interested in seeing whether there is some relationship between gender and political party affiliation. Your hypothesis is two-tailed, so you simply state that there is a relationship between the two, or using the null form, there is no relationship between gender and political party affiliation. It seems logical to see whether political party choice is an outcome of gender (clearly, choosing a party doesn't result in your gender); in other words, political party affiliation depends on the independent variable of gender. The values for gender (male and female) compose the columns, and the values for party (Republican, Democrat, Independent, Other) become the rows.

Table 7.1, a bivariate frequency table or cross-tabulation (crosstab), shows that there are 68 women who are Democrats in this sample (the cell at row 1, column 1). Remember, data are based on valid responses, and to be included, respondents had to answer both questions, with no blanks or missing values. Because percentage is used to standardize unequal category numbers, we can compare 69.4 percent with 37.7 percent in the next cell (row 1, column 2). Notice that although there are 68 women and 20 men who are Democrats, it would not be correct to say that more than three times as many women are Democrats than men. This would assume that the same number of men and women were answering the questions, but as you can

Table 7.1 Bivariate Cross-Table Example, SPSS

POLITICAL PARTY BY GENDER CROSS-TABULATION

			GENDER		
			Female	Male	Total
Political Party	Democrats	Count	68	20	88
		% of political party	77.3%	22.7%	100.0%
		% of gender	69.4%	37.7%	58.3%
	Republicans	Count	8	15	23
		% of political party	34.8%	65.2%	100.0%
		% of gender	8.2%	28.3%	15.2%
	Independents	Count	18	13	31
		% of political party	58.1%	41.9%	100.0%
		% of gender	18.4%	24.5%	20.5%
	Other	Count	4	5	9
		% of political party	44.4%	55.6%	100.0%
		% of gender	4.1%	9.4%	6.0%
Total		Count	98	53	151
		% of political party	64.9%	35.1%	100.0%
		% of gender	100.0%	100.0%	100.0%

see, the total sample is 98 women and 53 men. To make up for the unequal totals for each value of the independent variable, a percentage is used instead to make the comparison. It would be more accurate to say that approximately twice as many women as men are Democrats, in terms of the percentage that is relative to their sample sizes.

The total numbers at the bottom of the columns and at the right end of the rows are called *marginals,* because they appear at the edges or margins of the table. They essentially provide you the same information you would have gotten if you had constructed two separate univariate frequency tables. Note that this table includes percentages in terms of the dependent, or row, variable in addition to column percentages. For example, the 13 respondents who are male and registered as Independents (row 3, column 2) can be read either as 41.9 percent of the men are Independents, or you can say that 24.5 percent of all Independents in your study are male.

These are two different things; it all depends on how you want to report your data. If someone from the Republican Party wanted to know the breakdown of men and women in the party, then you would use the row percentages (percent of political party). If, on the other hand, you wanted to know whether there was a difference between men and women in terms of Republican affiliation, then use the column percentages (percent of gender). Sometimes there is no clear-cut independent (cause) and dependent (effect) variable because both can occur at the same time. These are often referred to as *symmetrical* relationships because either can be seen as the predictor for the other and either one can be the outcome needing explanation.

This can be confusing, so think aloud about what you really want to know and figure out how to show the percentages according to your objectives. It is not the same thing to say that 20 percent of women on campus major in political science and that 20 percent of political science majors are women. These are two different meanings, and it is crucial to make the distinction when presenting your findings in bivariate cross-tabulations.

A good rule of thumb is the values of a variable that you want to compare should add up to 100 percent. If you want to compare men with women and which majors they choose, the two categories (values) of gender should each add up to 100 percent. Then you can say what percentage of men select political science compared to what percentage of women do, and so on. But if you want to compare the majors and find out how many men and women are in each major, the categories for major (political science, sociology, psychology, etc.) should each add up to 100 percent. Then you can find out what percentage of political science majors are female and male compared to the percentage of sociology majors who are female and male, and so on.

TESTING BIVARIATE RELATIONSHIPS

Crosstabs illustrate whether some relationship is occurring with your data. There seems in Table 7.1 to be some difference in political party affiliation between men and women. Sometimes, however, the percentages and frequencies are so close that it is not possible to tell if there is a meaningful difference at a glance. Even when there appears to be a difference, you still need a more objective assessment than your own, perhaps selective, perspective to tell you whether the relationship is a significant one and not something that could occur just by chance.

Chi-Square

Rest assured, there is a way of testing whether there is a significant relationship between your variables! One of the most important and frequently used statistics to assess the association between ordinal and nominal measures is called chi-square (χ^2, pronounced "kie-square"). This measures how independent your two variables are and asks whether what you found (observed) is significantly different from what you would have expected to get by chance alone. The calculation looks at each cell and measures the difference between the actual frequency you got and the frequency that would have been expected by chance. For example, if half the sample is male and half are female, and half are Latino and half are Asian, then you would expect by chance 25 percent of the sample to be in each cell in that 2×2 table; that is, 25 percent should be Latino men, 25 percent should be Asian women, and so on.

Like the concept of the standard deviation, a mean deviation for each cell is calculated, and these are added together to produce a number called chi-square. Obviously, when there are many values for both variables, the larger the final number will be since there are many more cells to add together. This number is compared to a sampling distribution of chi-square values; the chi-square distribution approaches a normal curve when the number of cells increases, or more accurately when the degrees of freedom increase (see Box 7.1). The chi-square number in itself is not a measure of strength or magnitude; it must be compared to a distribution of chi-square values not to other chi-squares unless the sample sizes and number of cells are the same. Its value depends on the size of the sample and the number of cells: The more cells there are, the larger the chi-square value is likely to be.

The probability of obtaining a chi-square value by chance for a particular number of cells in a table is determined by the computer program or by comparing the value to a normal curve table of probabilities. If the probability is less than .05 (or whatever alpha level you set), then you can declare there is an association between your two variables by rejecting the null hypothesis of no association.

For the crosstab in Table 7.1 on political party and gender, a chi-square value of 17.361 was calculated for these data and is shown in Table 7.2. Notice there are three degrees of freedom (*df*) since there are four political party rows (4–1 = 3) and two gender columns (2–1 = 1), resulting in 3 × 1 = 3. What this statistical test tells us is that for a table with three degrees of freedom and 151 respondents, a chi-square value of 17.361 is significant at the .001 level for a two-tailed hypothesis: There is no difference in political parties between men and women. The probability of obtaining a chi-square value of 17.361 by chance alone is less than 1 in 1,000 ($p < .001$). We therefore reject the null, accept the alternative hypothesis that there is a relationship between gender and political party preference, and conclude that women in this sample are statistically more likely to be registered Democrats than are men. It's not enough just to say there is a relationship—you must also say what it is, or else the press conference will be fairly dull! Remember also that you are talking about a collection of men and women, not about any one particular person. Although you can say that women are more likely to be registered Democrats in this sample, you cannot generalize to all women (unless this is a random sample, which it is not), and you cannot say that any individual woman will be a registered Democrat.

The value of the chi-square statistic is also affected by cells with low frequencies, hence the information provided in footnote "a" that lets you know how many cells have small numbers. A rule of thumb is that every cell should have at least five expected respondents; you can see now how the size of the sample can make a difference when you have a table that is large, say five racial/ethnic groups and five religions, resulting in a 25-cell table. If your table has small numbers in its cells, there are statistical corrections, such as Fisher's Exact Test. If you have a 2 × 2 table, Yates' Correction should be used to adjust the calculation of chi-square. More information about these statistics can be found in advanced statistics books, on the Internet, and in most computer statistical programs.

Table 7.2   Example of a Chi-Square Test, SPSS

CHI-SQUARE TEST

	Value	*df*	Asymptotic Significance (2-tailed)
Pearson chi-square	17.361[a]	3	.001
N of valid cases	151		

[a] One cell (12.5 percent) has expected count less than 5. The minimum expected count is 3.16.

BOX 7.1
CALCULATING CHI-SQUARE AND DEGREES OF FREEDOM

$$\chi^2 = \sum \frac{(O-E)^2}{E}$$

Calculating a chi-square by hand is fairly simple. For each cell in the table, take what is actually found in the study or the observed value (O) and subtract from it the expected value (E). Figuring out what is expected in each cell is the tricky part. This is calculated in several steps: (1) Take the total number of people in a column and divide that by the total of all respondents answering the questions in order to get what percentage of the entire valid sample holds the value of the variable in the first column; (2) then use that percentage and multiply it by the total row value for each row associated with the first column. Next go on to the second column, take its total and divide by the grand total, and use that percentage and multiply it by the row total for each row in the second column, and so on. Using a formula, it's

$$\frac{\text{Column total}}{\text{Total total}} \times \text{Row total}$$

Imagine these are your results in raw numbers:

	MEN	WOMEN	TOTALS
Psychology	28	56	84
Sociology	26	35	61
Political science	67	27	94
Totals	121	118	239

1. Take the column total for men and divide by the total of all those answering the questions: 121/239 = 0.506 or 50.6%
2. Multiply that percentage by the row total for each row in the men column to get the expected:

50.6% × 84 = 42.5
50.6% × 61 = 30.9
50.6% × 94 = 47.6

Note these numbers add up to 121, the column total. All we did was to say that if men make up 50.6 percent of the total number of people answering these two questions, then we would expect 50.6 percent of psychology majors to be men, and 50.6 percent of sociology majors, and 50.6 percent of political science majors. Since there are 84 psych majors, we would expect 50.6 percent of 84 to be men, that is, 42.5, and so on for each major.

BOX 7.1 CONTINUED

In each cell, the expected frequency is then subtracted from the observed one, that number is squared, and the result is divided by the expected frequency. Do this for each cell, and sum all those numbers to arrive at chi-square.

In our example, the observed for the first cell is 28 and the expected is 42.5. Already we can see that there is a deviation from the expected. There is an underrepresentation of men who are psych majors.

1. First we subtract the expected from the observed:

   $28 - 42.5 = -14.5$

2. Then we square the difference; recall how we did something similar to get the standard deviation—we squared the deviations of each score from the mean, which is the expected:

   $-14.5^2 = (-14.5) \times (-14.5) = 210.25$

3. Next, we divide this number by the expected:

   $210.25 / 42.5 = 4.95$

4. We continue to do this for each cell:

   $26 - 30.9 = -4.9$

   $4.9^2 = 24.01$

   $24.01 / 30.9 = 0.78$ and so on for each cell.

5. We add up all the calculations for each cell to arrive at chi-square:

   $4.95 + 0.78 + 7.91 + 5.07 + 0.80 + 8.11 = 27.62$

The chi-square value is compared to a distribution of chi-squares to find the probability of achieving that chi-square by chance. Should it be less than .05 (or whatever standard was set such as .01 or .001), then you declare a statistically significant finding and hold your press conference!

Remember the size itself tells you nothing unless you know the number of cells. In this example, 27.62 is not necessarily statistically significant just because it is a particular value. A chi-square with a value of 8.5, for example, might be significant and one of 30 might not. It all depends on the number of cells and the size of the sample.

Degrees of Freedom

In this example, we have a 3 × 2 table, or six cells. Probability levels make use of a concept called degrees of freedom. *Degrees of freedom* (*df*) are the number of values that can vary in any calculation. When there is a fixed outcome (such as the sum of deviations around a mean always equals zero), it limits how many numbers can contribute to that outcome, usually by one ($N-1$). For example, consider the following: The total number of seats in a theater row was 10, and three groups wanted to be seated. A party of four enters, then a group of two; only four more people can now be accommodated. The first two numbers, four and two, could have been any size—they were free to vary. Once established, however, the next number is fixed. If a party of five came first, then three more came along, there is now only room for two. Out of three groups of people, two are free to vary, one is fixed, or $N-1$.

BOX 7.1 CONTINUED

For cells in a crosstab, the degrees of freedom are the number of rows minus one times the number of columns minus one or $(r-1) \times (c-1)$. In this example, it would be two rows (3−1) and one column (2−1), for a total of two degrees of freedom.

In fact, if you calculated the expected frequencies for the first cell (row 1, column 1) and then for the second cell in the example (row 2, column 1), you didn't have to calculate any more. All you had to do was subtract the expected in the first cell (42.5) from the row total of 84 to figure out the expected for the cell in row 1, column 2. It has to add up to 84, so the expected for that cell must be 41.5. That cell's value is not free to vary once you calculated the first cell's. Notice that with two numbers adding up to 84 one is fixed and one is free to vary ($df = N-1$).

Similarly, add up the expected frequency from cell 1 with the one below it (30.9) and subtract that number from the column total of 121 to get the expected value of the cell at row 3, column 1: $42.5 + 30.9 = 73.4$, then $121 - 73.4 = 47.6$. In short, only two cells are free to vary and had to be calculated; the rest are not free to vary. Again the degrees of freedom are $N-1$; in this case three rows minus one results in two. Out of six cells, there are two degrees of freedom and four are fixed. Degrees of freedom are also used as a correction when estimating a population parameter from a sample statistic, as was seen in Chapter 6 when discussing the formula to calculate the standard deviation with its denominator of $N-1$.

Correlation Coefficients

Chi-square is not the best statistic to use if you want to compare the results with other data because it is not a measure of comparative strength. If you are interested in the *strength* of an association, then correlation coefficients must be calculated. There are many available depending on the level of measurement for the variables being assessed. *Coefficients* range in value from 0.0 to 1.0. Like all coefficients, the closer a coefficient is to 1.0, the stronger the association. The closer a coefficient is to 0.0, the weaker the relationship. Except for nominal data where there is no order, coefficients also have either a negative or positive direction. A relationship is an *inverse* or *negative* one when an increase in one variable goes along with a decrease in the other; it's a *positive* relationship when a decrease in one variable goes along with a decrease in the other, or an increase in one goes with an increase in the other.

The strength of the correlation is not weakened in any way with the appearance of a minus sign. A correlation of −0.54 is stronger than a positive correlation of 0.45, for example. Sometimes the direction is due simply to the way the variables were measured; that is, if one variable is measured with 1 = strongly agree and 5 = strongly disagree, and another variable has 1 = strongly disagree and 5 = strongly agree, then a negative correlation between two variables measuring similar traits could result.

An easy way to understand coefficients is to think about them as the amount of change you have in your pocket. The closer the number is to $1.00, the more you have. If your coefficient is 0.83, then you have 83 cents, almost a dollar. If you get a coefficient of 0.17, then you have small change. Roughly, coefficients below 0.30 are considered weak, those between 0.30 and 0.70 are moderate, and those above 0.70 are fairly strong. But it's all relative in comparison to other studies and the size of the sample: If most research has found a coefficient of 0.14 between religious affiliation and political party choice, and you get 0.35, then yours might be considered fairly strong.

When evaluating a coefficient and its significance level, be sure to make a distinction about which numbers you are reading. Correlation coefficients range from 0.0 to 1.0, with the larger number meaning a strong and, very likely, statistically significant relationship. On the other hand, probability levels also range from 0.0 to 1.0, but smaller numbers mean low probability of chance occurrences. Because the goal is to get a statistic that did not happen by chance a good deal of the time—it is a finding due to the actual influence of the independent variable on the dependent variable, not due to chance—then the objective is to have a probability level that is low: $p < .05$, $p < .01$, or $p < .001$, the typical numbers used as cutoff points for significance levels. The closer the probability level is to 0 (that is, no probability the statistic occurred by chance), the more likely that what you found did not occur accidentally. The closer the correlation coefficient is to 1.0, the stronger it is; the independent variable explains almost all the variation in the dependent variable. You will make fewer errors in explaining or predicting it.

It is important to know, though, that low correlation coefficients can be statistically significant with large sample sizes; thus, it is better to look at the strength of correlations rather than to focus only on their significance levels.

Measures of Association: Nominal Variables

Although chi-square is the statistic most used to test the association between two variables, many other statistics are available to assess the strength of a relationship. When you are interested in evaluating relationships between two *nominal* level variables, then use such statistics as the *Phi* (φ) *coefficient* (for 2 × 2 tables), the *Contingency coefficient* (for larger than 2 × 2 tables), *Cramer's V* (for tables without the same number of columns and rows), *Goodman and Kruskal's Tau* (T), and *Lambda* (λ). These are discussed in more detail in advanced statistics books and online resources. For our purposes, the goal is to understand how to interpret correlation coefficients.

Consider again the results between political party and gender found in Table 7.1. Note that the Phi, Cramer's V, and Contingency coefficients are all fairly similar in Table 7.3, around 0.32 to 0.34 (rounding off), and all are statistically significant ($p < .001$). Chi-square told us that there is a relationship, and these statistics communicate that it's a moderate one in strength. Chi-square does not let us know how strong or weak the relationship is, only that there is one.

Table 7.3 Example of Statistics for Nominal Variables, SPSS

SYMMETRIC MEASURES

		Value	Approximate Significance
Nominal measures	Phi	0.339	.001
	Cramer's V	0.339	.001
	Contingency coefficient	0.321	.001
N of valid cases		151	

Lambda and Tau. Many researchers prefer to use coefficients that determine the *proportional reduction in error* (PRE). *Lambda* and *Goodman and Kruskal's Tau* are two statistics best suited for this when you have nominal variables. What PRE means is that the values of Lambda and Tau tell us approximately how many fewer errors we will make when predicting the outcome values of the dependent variable once we know the values of the independent variable. Be aware, though, that Lambda is based on modes, so it is possible to obtain a Lambda of 0. This simply means that a prediction of the dependent variable mode is not helped by knowing the modes of each category of the independent variable.

Because it is calculated using modes, the value of Lambda is a function of which variable is the dependent one. Note that several values are calculated, depending on whether the association you are assessing is symmetric or asymmetric. *Asymmetric* (or directional) measures assume that one of the variables is definitely dependent on the other; *symmetric* measures allow for the possibility of either one being dependent. When your hypothesis is one-directional, then asymmetric Lambda or Tau is more appropriate. If you are interested, for example, in the relationship between gender and race/ethnicity in your sample, then you would look at symmetric Lambda because neither variable occurs before the other in time or is being used to predict the other.

Because political party is the dependent variable, the value of Lambda is 0.000, the second one in Table 7.4. According to these statistics, there is no correlation between gender and political party, hence there is no significance level given. With most of the output, focus primarily on two items: the value of the statistic and its significance level. The other information provided is useful for those with more advanced statistical training.

Lambda is calculated using the mode; the modal political party for men in this sample is Democrat, and the modal political party for women is also Democrat. Knowing respondents' genders does not improve or reduce our errors when predicting their political preference. Your best guess in either case is Democrat, regard-

Table 7.4 Example of Lambda and Tau Statistics for Nominal Variables, SPSS

DIRECTIONAL MEASURES

			Value	Asymp. Std. Error[a]	Approx. T[b]	Approx. Signif.
Nominal measures	Lambda	Symmetric	0.069	.046	1.424	.155
		Political party dependent	0.000	.000		
		Gender dependent	0.151	.098	1.424	.155
	Goodman and Kruskal's Tau	Political party dependent	0.057	.028		.000[c]
		Gender dependent	0.115	.053		.001[c]

[a] Not assuming the null hypothesis.

[b] Using the asymptotic standard error assuming the null hypothesis.

[c] Based on chi-square approximation.

less of the gender. And that is your best guess in general because the mode for the overall sample is Democrat as well.

Here is how you interpret a PRE if there is some correlation coefficient greater than 0. Let's say you get a value for Lambda of 0.45 between gender and religious affiliation. This coefficient tells you that errors in predicting the religion of the respondents in your sample are reduced by a proportion of 0.45 (or 45 percent) when you know what their genders are. Imagine you and your friends go to a "Guess Your Occupation" carnival booth. Suppose the person guessing your jobs knew nothing about you; you hide behind a screen and don't even speak. The guesses will be all over the place because there are no hints or clues to help the guesser. So you help out, and you and your friends disclose your genders. There will be fewer errors now since many men tend to have different kinds of jobs than many women. By how much does the guesser improve? The PRE tells you that. So, for nominal level data, Lambda and Tau provide you with this information. Not only do you get a sense of the strength of the relationship and can compare the results to other relationship coefficients, you also know how well you will be reducing errors in predicting (PRE) and explaining the dependent variable.

Because the goals of research include explaining and predicting as much as we can about our outcome variables, it is important to calculate statistics that determine how well we are doing with our independent variables. Chi-square tells us whether our two variables are independent of each other, or to put it another way, whether they are associated. Once we know that they are related, coefficient measures like Tau and Lambda tell us how strongly the variables are associated and how much error reduction we have when we predict the dependent variable values knowing the independent variable values.

Measures of Association: Ordinal Variables

In addition to chi-square, several other statistics are appropriate when the categories of the variables are ordered. If you wish to assess the association of two ordinal variables, such statistics as Gamma (γ), Kendall's Tau-b and Tau-c, and Somers' d can be used to determine the strength of the association. These coefficients also range from 0.0 to 1.0 but, unlike the ones for nominal data, they can be negative or positive, depending on the order of the categories. Each of these statistics has sampling distributions that are used to determine the probability of obtaining that statistic by chance. So, if the probability of obtaining the value calculated for Gamma or Kendall's Tau-c by chance is less than the level you set (minimally .05), then you reject the null hypothesis and declare there is an association between the independent and dependent variables.

Gamma. A popular statistic to use is *Gamma,* which is based on a concept of evaluating pairs of responses and whether the respondents' relative order on both the independent and dependent variables is similar or dissimilar. Gamma is a symmetric statistic (only one coefficient is calculated because it doesn't matter which is independent or dependent) and a PRE measure. This means that the coefficient itself not only tells you the strength of the relationship but it also indicates the proportion of reducing error in predicting the dependent variable once you have information on the independent variable. See Table 7.5, which shows the results of a study that found a Gamma of 0.191 between two ordinal variables: political views (liberal, moderate, conservative) and attending religious services (several times a year or less versus once a month or more). Although statistically significant, this could be interpreted as a weak relationship in which 19 percent of the errors in predicting religious attendance are reduced by knowing respondents' political views.

Kendall's Tau. *Kendall's Tau-b* (not to be confused with Goodman and Kruskal's Tau for nominal level variables) uses different methods for calculation than Gamma and assumes there are the same number of rows and columns in the table. *Kendall's Tau-c* can be used for tables of data in which the numbers of rows and columns are not equal. In this example, because there are three categories for political views and

Table 7.5 Example of Kendall's Tau and Gamma Statistics for Ordinal Variables, SPSS

SYMMETRIC MEASURES

		Value	Asymp. Std. Error[a]	Approx. T[b]	Approx. Signif.
Ordinal by ordinal	Kendall's Tau-b	0.111	0.026	4.298	.000
	Kendall's Tau-c	0.127	0.030	4.298	.000
	Gamma	0.191	0.044	4.298	.000
N of valid cases		1,316			

[a] Not assuming the null hypothesis.

[b] Using the asymptotic standard error assuming the null hypothesis.

two for religious attendance, Tau-c could be used. These are symmetric statistics, and the same coefficient of 0.127 would result if we decided political views depended on religious attendance. But Kendall's statistics are not PRE measures. The coefficient calculated can only be interpreted in terms of strength (weak to strong) and direction (positive or negative/inverse relationship). In this case, it is positive but weak.

Somers' d. Similar to Gamma in the way it is calculated (that is, based on comparing the similarity or dissimilarity of ranking on pairs of responses on the two variables), *Somers' d* differs in that it can be an asymmetric statistic where two values are calculated, depending on which is the dependent variable. However, it does not have a PRE interpretation.

In Table 7.6, the correlation between political views (in order: liberal, moderate, conservative) and attending religious services (several times a year or less versus once a month or more) is 0.110 when you assume it is symmetrical and one variable is not predicting the other. If you feel that political views precede religious attendance, then the value of Somers' d is 0.096. In either case, it is statistically significant ($p < .001$) but a weak positive coefficient. This means there is some relationship between those who are conservative (a high value on the political views measure) and attending services once a month or more (a high value on the religious services measure), or conversely, those who are more liberal in this sample seem to attend religious services less frequently than those who are conservative. It is statistically significant at such a low level of strength because it is from a large sample, in this case, a sample of over 1,300 respondents.

The differences among all these ordinal statistics are the way they deal differently with pair rankings that are tied, something discussed in more detail in advanced

BOX 7.2
CORRELATIONS IN PUBLISHED RESEARCH

Here are two excerpts from articles in academic publications.

1. Baumann (2001) studied audiences' shift in perceptions about movies from a form of entertainment to cultural art. One argument is that seeing film as an art form was partly due to changes in how critics wrote about movies over time. Consider these findings (adapted from Baumann 2001: 414):

NUMBER OF "HIGH ART" AND "CRITICAL" TERMS USED IN TWELVE FILM REVIEWS FOR EACH YEAR FROM THE *NEW YORK TIMES, NEW YORKER,* **AND** *TIME* **MAGAZINE**

Number	Year
19	1925
18	1930
21	1935
5	1940
21	1945
9	1950
12	1955
31	1960
32	1965
107	1970
69	1975
103	1980
78	1985

The researcher tests the one-directional hypothesis that over time there is an increase in the number of artistic and critical words used in film reviews. Since year is already in rank order, the author decided to use Spearman's rho to test the correlation. This requires ranking the dependent variable (number of artistic words), and the rank order for artistic words is then compared with the rank order of years. The results are rho $= 0.88$, $p < .001$ (one-tailed test). This is a very strong positive correlation, leading to the conclusion that in the late 1950s through the late 1960s, movies became valorized as an art form. Year of the review explains more than 77 percent of the ranking variation in the number of artistic and critical words. (Spearman's rho is a PRE measure that must be squared to calculate the error reduction.)

2. Huhman et al. (2010) studied exposure to a national mass media campaign encouraging physical exercise and activities and its impact on children's physical activity, along with some psychosocial variables like peer influence and self-efficacy in trying new activities. Here are some of the findings looking at the relationship between media exposure to the campaign and whether the children (ages 10 to 13) said they engaged in physical activity the day before the data were collected (adapted from Huhman et al. 2010: 642).

BOX 7.2 CONTINUED

	CAMPAIGN EXPOSURE (PERCENTAGE)				
	None	Less Than Once a Week	About Once a Week	Several Times per Week	Every Day
Engaged in physical activity yesterday	62.4	63.4	64.2	70.5	68.4
Did not engage in physical activity yesterday	37.6	36.6	35.8	29.5	31.6

$\gamma = 0.09, p < .05$

To test the strength of the relationship, the authors used Gamma (γ) because the independent variable (exposure to the media campaign) is an ordinal measure, and the dependent variable (physical activity) is a dichotomy that can be treated as ordinal. Although the Gamma is statistically significant ($p < .05$), it is a weak correlation. The PRE tells us that errors in predicting physical activity are reduced only by 9 percent when the amount of media exposure the children experienced is known. Yet, it can be seen that more of those who saw the physical campaign materials engaged in physical activities.

statistics books or online resources. Yet, notice how similar they are in terms of the actual coefficient values. However, if you want a statistic that not only provides a measure of the strength of the association but also indicates how much error reduction occurs, then Gamma is ideal for ordinal measures.

Table 7.6   Example of Somers' d Statistic for Ordinal Variables, SPSS

			DIRECTIONAL MEASURES		
		Value	Asymp. Std. Error[a]	Approx. T[b]	Approx. Signif.
Ordinal by ordinal	Somers' d Symmetric	0.110	0.025	4.298	.000
	Religious services attendance dependent	0.096	0.022	4.298	.000
	Political views dependent	0.128	0.030	4.298	.000

[a] Not assuming the null hypothesis.

[b] Using the asymptotic standard error assuming the null hypothesis.

Spearman's rho. Sometimes both the independent and dependent variables consist of rank ordered numbers. For example, you want to compare the rank order of people in terms of their grades on the first research methods exam with their rank order on the last exam taken. So, student A might be ranked third on the first test and fifth on the last test, student B might be twelfth on the first test and second on the last test, and so on for each of the 100 students in your study. Note that you are comparing not actual grades (which are interval/ratio measures) but rather their rank order number. In such cases, *Spearman's rank order correlation coefficient* or *rho* (ρ) provides a measure of association.

Spearman's rho is a symmetrical statistic that results in a coefficient between 0.0 and 1.0 to indicate the strength of the relationship and a plus or minus to show the direction of the relationship. In addition, it's a PRE measure when you square the value of rho. For example, if the correlation coefficient is 0.65 between the first set of grades and the last set, then you can conclude that there is a strong positive relationship between the two and that error in predicting rank order on the last exam is reduced by around 42 percent (0.65 squared). If for some reason the coefficient were −0.65, then you would conclude that those who rank higher on the first exam tend to rank lower on the last exam, and those who rank lower on the first rank higher on the last.

Measures of Association: Interval/Ratio Variables

Ranking scores on a test, as in the previous example, loses information. The person with the best grade could be only 1 point higher than the next person or 20 points higher. Ordinal measures also do not give you enough information like the distance between the ranks. Certainly, if you have interval/ratio measures, then make use of them, because they represent the best level of measurement for the more sophisticated statistical analyses. There are several ways you can evaluate a relationship between interval/ratio variables.

Scatterplots. Although you are not likely to put interval/ratio variables into cross-tabulations when there are a large number of values for each variable, you can visually assess the relationship with a graph. Most common are *scatterplots*. As with frequency curves for univariate data, scatterplots assume interval/ratio measures. The horizontal x-axis is usually for the values of the independent variable, and the vertical y-axis for the units of the dependent variable. A mark is placed at the point where each surveyed person's responses to the two variables intersect, not unlike the idea of a cell in a contingency table for nominal and ordinal variables. When all the responses are plotted, a pattern emerges.

When the points are scattered everywhere in the graph with no apparent pattern, it is signifying no relationship, that is, an association close to 0.0 coefficient. When the points tend to fall along an almost straight line, then there is a perfect relationship close to 1.0. If it looks like an uphill line, then it's a positive linear relationship; if it looks like a downhill line, it's a negative/inverse linear one. When the points tend to form a curve, then it's called a curvilinear relationship.

Take, for example, the relationship between height and weight. Usually, taller people weigh more, and shorter people weigh less. But as we all know, this is not a perfect guide; there are many short heavy people and tall thin people. We would expect then a scatterplot that looks like an uphill slope, but not all the points would fall in a perfect straight line.

As you can see from the scatterplot in Figure 7.2, there is a tendency for those who are shorter (the x-axis marks the height in inches) to be lighter in weight (the y-axis marks the weight in pounds). Each point on the chart represents one person's height and weight. If you were to draw an exact straight line from any point to the x-axis you would find out the height for that respondent, and a line from the same point to the y-axis would indicate the weight. Notice that there are some outliers, that is, respondents who fall beyond an imaginary straight line that could be drawn through the middle of the points. For example, there appears to be three people around 60 inches tall but one of the points is closer to 175 pounds or so on the y-axis. That person's data falls outside the general pattern.

Pearson r. Although scatterplots are not as precise as a table of numbers, they do provide a rough visual indication of the kind of relationship that exists between two variables. However, for a more specific assessment of the magnitude (weak to

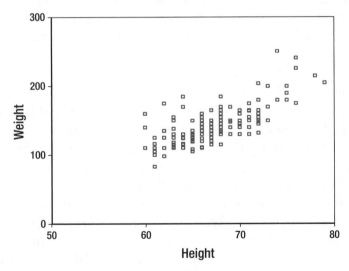

Figure 7.2   Example of a Scatterplot, SPSS

strong) and direction of the relationship (positive or negative/inverse), *Pearson's product-moment correlation coefficient,* a symmetric statistic known simply as *Pearson r,* is calculated. Similar to the other coefficients discussed, it has two components: a number ranging anywhere from 0.0 to 1.0 to indicate strength, and a plus or minus sign to show direction. Like Gamma, Lambda, and Goodman and Kruskal's Tau, Pearson r is also a PRE measure, but like Spearman rho, it must be squared to establish the proportion of error reduction.

Pearson r is one of the most used statistics in social science research and central to path analysis, linear regression analysis, and other statistical methods. Its calculation is based on means, standard deviations, and z-scores. In order to compare variables with two different units of measurement, such as height and weight, respondents' answers are transformed into z-scores (as described in Chapter 6). What Pearson r does is measure how much change in the z-scores of one variable is related to change in the z-scores of the other variable. Is the variation from respondent A's z-score for height to that of respondent B's similar to the variation from respondent A's z-score for weight to respondent B's, and so on throughout the sample of all respondents? Every change in an independent variable z-score (a standard deviation unit) is related to a change in a dependent variable z-score. That is why it is called a correlation. If those changes vary at the same rate and with the same incremental units, then there is a perfect 1.0 correlation coefficient that looks like a straight line in a scatterplot, either going up if positive (/) or going down if it's an inverse relationship (\) .

The data correlating height and weight result in a Pearson r of 0.714 (Table 7.7). For every change of one standard z-score unit in height (independent variable) there is a 0.714 standard unit change on the outcome or dependent variable (weight) in the same direction. Those who are taller tend to weigh more, and those who are shorter tend to weigh less. If it were an inverse relationship of −0.714, then we would conclude that taller people weigh less and shorter people weigh more. The correlation does not tell you about any one person, so it is not accurate to say that any one short person weighs less. These are aggregated data and are only interpretable in terms of the entire sample, not in terms of one person at a time. Because the goal is to predict variation across the sample on the dependent variable, you can also say that you have reduced errors in predicting the weights of all respondents by 51 percent (r^2), by knowing information about their heights.

The properties of this statistic also allow you to say it another way: You have determined 51 percent of the variation in weight (the dependent variable) through your knowledge of height (the independent variable). This is why r^2 is called the *coefficient of determination.* In this particular example, of all the variation that exists in your sample for weight (after all, not everyone is the same weight), you account for

Table 7.7 Example of Pearson r Statistic for Interval/Ratio Variables, SPSS

CORRELATIONS

		Height	Weight
Height	Pearson correlation	1	0.714*
	Significance (2-tailed)		.000
	N	151	149
Weight	Pearson correlation	0.714*	1
	Significance (2-tailed)	.000	
	N	149	150

*Correlation is significant at the .01 level.

51 percent of it with the respondents' heights. Height is only one cause or explanation or predictor, albeit a major one, of the outcome, namely, weight. The remaining 49 percent of the variation (sometimes referred to as the *residual*) must be explained or predicted by other reasons such as genetics, food habits, and amount of exercise.

As with the other statistics, Pearson r has a sampling distribution, and we can use it to ask what the probability of obtaining a particular coefficient is by chance alone. If the probability is less than .05, we reject the null hypothesis and declare there is a statistically significant association between the independent and dependent variables. However, when sample sizes are large, small coefficients tend to be statistically significant, and for this reason, many researchers look primarily to the strength of the relationship and not as much to its significance level.

Eta. Pearson r also requires that the relationship between the two variables is linear; that is, it is not a curve. As one variable increases or decreases, so too does the other. An example of a nonlinear relationship might be age and muscle strength. As you grow older, you get stronger, but then after a certain age, your muscles get weaker. Strength does not continue to grow linearly as you get older. Pearson r would not be an adequate measure of the relationship between age and physical strength. It would be more appropriate to use *eta* (η), which is designed for nonlinear associations. It requires an interval/ratio dependent variable, and the independent variable should be in nominal or ordinal categories. When squared, eta is a PRE measure and can be interpreted similarly to r^2. So, if you find an eta of 0.50 between age and muscle strength, then you can say that you can reduce your errors in predicting physical strength by 25 percent by knowing different categories of age, such as younger people are stronger and older people are weaker.

The Purposes of Measuring Relationships

Remember that the main goals of doing research are to describe, explain, or predict. The objective is to account for the reasons why the dependent variable varies among the respondents in your study, to predict future occurrences, or simply to describe any relationships among your variables. Measures of association are a good way of evaluating these relationships, to see if they exist at all and, if so, how strong they are and in what direction they go. It also becomes necessary sometimes to elaborate on the original relationship, to control for alternative explanations, and to test for spurious relationships. Chapter 9 discusses how to elaborate your findings by introducing a third or control variable into the analyses.

When we analyze relationships and generate PRE information, we answer some of the research questions or hypotheses we began with, thereby contributing to theory building, to the assessment and evaluation of programs, and to descriptions of some social phenomena. However, sometimes we are interested in describing, explaining, or predicting differences among various categories of people and not just uncovering associations between variables. Ways of analyzing data for differences are discussed in the next chapter.

REVIEW: WHAT DO THESE KEY TERMS MEAN?

Bivariate analysis	Eta	Pearson r
Cells	Gamma	PRE
Chi-square	Goodman and	Residual
Coefficient	Kruskal's Tau	Rows and columns
Coefficient of	Kendall's Tau-b and	Scatterplots
determination	Tau-c	Spearman's rho
Contingency tables	Lambda	Symmetric and
Cross-tabulations	Marginals	asymmetric
Degrees of freedom	Outliers	

TEST YOURSELF

Here are several hypotheses: First identify independent and dependent variables and their levels of measurement. Then say which statistic would be most useful to test whether you reject or accept the null hypothesis.

1. There is no relationship between SAT scores and college GPA.

	Which Variable?	Level of Measurement?	Which Statistic to Use?
Independent variable			
Dependent variable			

2. There is no relationship between type of car owned and region of the country.

	Which Variable?	Level of Measurement?	Which Statistic to Use?
Independent variable			
Dependent variable			

3. There is no relationship between the Top 20 college football rankings this year compared to these schools' rankings last year.

	Which Variable?	Level of Measurement?	Which Statistic to Use?
Independent variable			
Dependent variable			

4. There is no relationship between gender and number of times using Twitter per day (measured as "none," "1 to 5 times," "6 to 10 times," "11 or more times").

	Which Variable?	Level of Measurement?	Which Statistic to Use?
Independent variable			
Dependent variable			

INTERPRET: WHAT DO THESE REAL EXAMPLES TELL US?

1. Table 7.8 shows some SPSS output from the 2010 General Social Survey.
   a. Which is the strongest correlation in this matrix? Note that the table is symmetrical because the correlation of "Hours per Day Watching TV" with "Highest Year of School Completed" is the same as "Highest Year of School Completed" with "Hours per Day Watching TV" and that correlations of variables with themselves always equal a perfect 1.0 correlation.

Table 7.8 Pearson r Correlations

		Highest Year of School Completed	Hours per Day Watching TV	Age of Respondent
		CORRELATIONS		
Highest year of school completed	Pearson correlation	1	−0.248**	−0.049*
	Significance (2-tailed)		.000	.027
	N	2,039	1,421	2,036
Hours per day watching TV	Pearson correlation	−0.248**	1	0.128**
	Significance (2-tailed)	.000		.000
	N	1,421	1,426	1,423
Age of respondent	Pearson correlation	−0.049*	0.128**	1
	Significance (2-tailed)	.027	.000	
	N	2,036	1,423	2,041

*Correlation is significant at the .05 level (2-tailed).

**Correlation is significant at the .01 level (2-tailed).

    b. How would you put into words the relationship between these two variables of education and TV viewing? What does the minus sign tell you?

    c. The correlation between age and years of schooling completed is statistically significant. Yet over 2,000 respondents answered these questions. How would you interpret this significance level in comparison to what seems to be a weak correlation?

2. Hansen and Mitchell (2000) published results of their study on the political activity of Fortune 500 corporations. They compared companies that have political action committees (PACs) that make donations to election candidates with those that don't on several variables, including whether they engage in lobbying the government with consultants and how much they give in charitable contributions. Here are a few results (2000: 896):

Table (A)	No PAC	PAC
No lobby	67%	25%
Lobby	33%	75%

$\chi^2 = 97.95$, $df = 1$, $p < .001$

Table (B)	No PAC	PAC
No charity	80%	45%
Charity	20%	55%

$\chi^2 = 75.32$, $df = 1$, $p < .001$

a. State the hypothesis being tested in each of these tables.
b. Put into words what is going on in each of these two tables. How do you read the tables? For example, 67 percent of what is what, and so on.
c. What does the chi-square tell you? How was *df* determined? What does the *p* value mean?
d. What do you conclude?
e. Which other statistics could you use to assess the strength of the relationship?

CONSULT: WHAT COULD BE DONE?

You are hired to do a marketing research analysis for the local newspaper. The publishers want to find out which kinds of people use the Internet to read the online version of the paper instead of buying the print version, how often, and which sections and features are most read and liked.

1. Preliminary data analysis suggests that men and women read different sections of the online paper. What would you do to see if this is statistically significant?
2. Frequency of reading the online version of the paper seems to vary based on such characteristics as income, educational level, age, and size of family. How would you statistically test each of these possible relationships (income and frequency of reading the paper, education and frequency, age and frequency, and size and frequency)?
3. What other relationships would you recommend be studied to provide the newspaper owners more information about readership?

DECIDE: WHAT DO YOU DO NEXT?

For your study on how people develop and maintain friendships, as well as the differences and similarities among diverse people, respond to the following items:

1. Review your bivariate hypotheses (or write a few new ones) and list the variables.
2. Describe the statistics you would use to evaluate the relationships and why you would use these.
3. If you have actual data, begin to analyze the hypotheses that are best suited for statistical measures of association and correlation.

8 ANALYZING DATA

Comparing Means

I note the obvious differences between each sort and type, but we are
more alike, my friends, than we are unalike.

—*Maya Angelou, poet*

LEARNING GOALS

This chapter shows you how to assess differences between means using t-tests and analysis of variance. As
with other bivariate data analysis, knowing when to use these statistical procedures and how to interpret
them is central to testing hypotheses. By the end of the chapter, you should be able to understand what
t-tests and ANOVA are and when they are suitable for data analysis.

Figure 8.1   Statistical Decision Steps

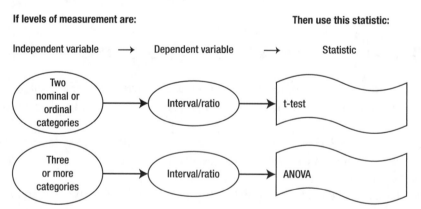

Every day it seems people are making comparisons among groups of people.
Women are from Venus, and men are from Mars, or something like that.
Young people drive differently than the elderly, African Americans don't

vote for the same candidates as whites, and certain cities have higher crime rates than others.

One of the fundamental questions asked in research is how alike or different people are. Good questionnaires ask demographic information that can provide important ways of breaking down survey data into interesting comparison categories. For example, you might be curious about the differences in the amount of television viewing among racial/ethnic groups or comparisons between people under 30 and people over 60 in the number of text messages they send per day.

If the data you want to compare are interval/ratio measures or at least ordinal measures with equal-appearing intervals, then there are several basic statistical techniques that can be used to answer the question of meaningful differences. For example, when you compare the average number of hours studied per week between first-year students and seniors, it is unlikely that the means will be exactly the same for both groups. Hence, the question is whether the difference that exists in hours studied between first-year students and seniors is a statistically significant one, as opposed to just any difference that could easily occur by chance alone.

When seeking to evaluate differences in means, we can use two very important and popular statistics: the *t-test* and *analysis of variance* (ANOVA).

T-TESTS

When what you want to know (the dependent variable) is a ratio/interval measure, an ordinal measure with equal-appearing intervals, or a dichotomy, and a mean and standard deviation can be calculated for this dependent variable for each of two categories of comparison (often two categories of the independent variable), you can select a *t-test* for data analysis. T-tests ask whether the difference between two means is significantly different from zero. If both means were identical, then subtracting one from the other would result in zero. When data come from a sample of respondents, you are technically asking whether the difference in the true means in the populations from which these samples are derived is zero. To put it another way, you are asking a null question: There is no difference between the two means in the sample; they are the same and are from the same population.

Remember, you are comparing information that you collected from just one sample of respondents. If all possible samples of a certain size were taken from the population and all possible differences in means between two categories for all those samples were calculated (for example, mean number of texts sent per day by urban people versus rural people) and plotted on a graph, the figure would approximate a normal distribution, as discussed in earlier chapters. Actually, they would form a t-distribution, slightly different in shape from a perfect normal curve, or so

"Student"—the nom de plume for W. S. Gosset, an employee in Ireland's Guinness brewery who developed the "Student's t"—realized.

What we are asking is the core question of statistical research in the social sciences: How much do our sample means deviate from the norm? Is the difference in means we found between two categories in our sample of respondents significantly different from zero? Where on the distribution of all possible mean differences is our mean difference? If it is 1.96 standard deviation units away (actually called *the standard error of the differences between means* when discussing distributions of mean differences), then you can conclude that the results are significant at the .05 level using a two-tailed or two-directional hypothesis. If we wanted to see whether one of the means was significantly larger than the other, or conversely one was smaller than the other, we are asking a one-directional question, and we would see whether the mean difference was 2.58 units away from the norm of zero difference in order to be significant at the .05 level.

Let's take an example in which we want to learn whether older or younger people download more movies onto their computer tablets in a typical month. First, we have to create two categories in order to make the comparison suitable for a t-test when there does not already exist a dichotomy. In this case, what is probably an interval/ratio measure (age) needs to be made into a simple dichotomy by deciding what constitutes "younger" and "older." We can look at the frequency distribution and cut the sample at the median, so that half the respondents are now above the median (older) and half are below (younger). Or we can decide ahead of time, based on previous research or theory, and make 40 or under the "younger" respondents and those 41 and above the "older" people, even though we may not get a 50 percent split this way.

Now we have two groups, and we can calculate the average number of movies downloaded in a typical month. Remember, the questionnaire item has to be written as an interval/ratio measure, or an ordinal one with equal-appearing intervals, so we can accurately calculate a mean and standard deviation. We hypothesize that the difference in average number of downloaded movies between the younger and older people is zero, and assuming we have no preconceived idea about which group is more likely to be heavier downloaders, we proceed with a two-directional t-test to assess the differences in means. If we assume, based on previous research, that younger people are more likely to download movies to their computer, then we will have a one-tailed hypothesis. We will definitely get some difference in means because it is unlikely that the two groups download exactly the same average number of movies, but we want to know if that difference is *statistically significant.*

The computer calculates a t-test value and presents us with the significance level. If the probability of obtaining that t-value by chance alone is less than .05, then we

BOX 8.1
EXPLAINING THE T-TEST

Just as we can tell how far away an individual score is from the group mean in terms of standard deviation units (that is, z is a score minus the mean divided by the standard deviation), so too t is a mean minus another mean divided by the standard error of the differences in means.

$$t = \frac{(\bar{x}_1 - \bar{x}_2) - (\mu_1 - \mu_2)}{s_{\bar{x}_1 - \bar{x}_2}}$$

In this formula, the two sample means are subtracted to find a difference, and the two population means are subtracted to find their difference. But because the null hypothesis is that the two population means are the same, the difference between them is zero. Hence, the second part of the numerator (where the μs are being subtracted) becomes zero and is not used in the calculations. The formula essentially is asking how different the difference in the two sample means is from zero, but all you need to know are the two sample means. It doesn't really matter what the real population means are since they are hypothesized to be the same, and therefore equal to zero when subtracted. Besides, if we knew what they were, we wouldn't be doing a sample survey in the first place!

The denominator is the *standard error of the differences between means*, and it is calculated differently depending on whether the variances of the two samples are the same or different and whether the data come from paired samples and how correlated they are, that is, the same people measured twice. Advanced statistics books and websites provide the formulas for the denominators. The main thing to remember is that the t is a ratio of the difference between two means (the numerator) divided by the standard error (the denominator).

When calculating the t-value for the difference between a sample mean and some known population mean, the formula is similarly a ratio of the mean differences divided by the standard error or deviation:

$$t = \frac{\bar{X} - \mu_0}{s_x}$$

conclude that the category with the higher mean downloads more movies per month on average than the group with the lower mean. The difference between these two means is declared to be significantly different from zero; that is, if we wrote a hypothesis in the null form, we would reject no difference between the two groups and hold a press conference to declare a significant finding. Rather than use the double negative of rejecting a null, we can more easily say that we are accepting a finding of difference between the two categories and then state which group uses their tablets to watch movies more or less than the other group.

Understanding T-Tests

When I was a kid, I often got frustrated when I did not reach the special line on the sign at the entrance to a really cool ride at the amusement park. I was too short to get on. I was sure I was shorter than everyone else in the world. But I guess I did not have it as bad as my sister and all her girlfriends, who were shorter than I. After all, we all know that girls are shorter than boys! Or do we? Let's take a look at this and see if indeed there is a significant difference between the average height of women and the average height of men in a survey of 151 college students.

Because height is an interval/ratio measure, you can calculate a mean and standard deviation for each category. And because you are comparing means between two groups within your sample, the t-test is the most appropriate statistic to use. You are testing to see if the difference between the two means is significantly different from zero. Let's just assume for the moment that you do not have any preconceived ideas about which group is taller or shorter. You are just interested in seeing if there is a difference and do not have any real interest in which group is taller or shorter. Therefore, you employ a two-directional test of significance.

Independent Samples T-Test

When the two subsamples are independent of one another—that is, it is impossible that someone could select both categories of the questionnaire item—we use the *independent samples t-test*. The first part of the SPSS output in Table 8.1 indicates that the average height of the 97 women who answered the question is 65.39 inches (approximately 5'5") and 70.72 inches (approximately 5'11") for the 54 men. There definitely is a difference in height, but is it a statistically significant difference? What is the probability that this difference could have happened by chance and not be a result of real differences between men and women? Is five or six inches a meaningful difference, given the way the samples of men and women are distributed and the size of the samples? If the male and female respondents each are widely dispersed, smaller differences will be less meaningful. If the distributions are narrowly dispersed—that is, there is a lower standard deviation—then a small difference will be much more meaningful.

The standard deviations in the first part of Table 8.1 tell us that the heights for the men in the study are more dispersed than those for the women (3.36 versus 2.94) and that the distribution of all possible means for men is wider than it is for women (the standard error of the means shows that). The calculation of the t-value is affected by these distributions.

The SPSS output for independent samples includes the results of the calculations for two different t ratios, depending on which standard error formula is used in the denominator. To determine which t-test to look at, you first must take a side trip

Table 8.1   Example of Independent Samples T-Test, SPSS

GROUP STATISTICS

	Gender	N	Mean	Standard Deviation	Standard Error Mean
Height	Female	97	65.3918	2.9353	0.2980
	Male	54	70.7222	3.3614	0.4574

INDEPENDENT SAMPLES TEST

	LEVENE'S TEST FOR EQUALITY OF VARIANCES		T-TEST FOR EQUALITY OF MEANS					95% CONFIDENCE INTERVAL OF THE MEAN	
	F	Significance	t	df	Significance (2-tailed)	Mean Difference	Standard Error Difference	Lower	Upper
Height									
Equal variances assumed	1.148	0.286	−10.148	149	.000	−5.3305	0.5252	−6.3684	−4.2946
Equal variances not assumed			−9.764	97.817	.000	−5.3305	0.5459	−6.4139	−4.2470

and look at the Levene F-test for equality of variance (hereafter called the F-value). The second part of Table 8.1 shows whether the two variances (that is, the standard deviation squared) are equal to one another. If the F-value calculated is significant ($p < .05$), then you conclude there is a difference and reject the null hypothesis of no difference in variances; equal variances are not assumed. If the significance level for the F-value is not significant ($p > .05$) then you accept the fact there is no difference and assume equal variances between the two categories. In the second part of the table, you can see that the significance level for the Levene F-value is greater than .05 (it is .286), so you assume that the variance of height for our sample of men and the variance of height for women are approximately equal.

But you're not done yet. You just took a side trip and finished figuring out if there was a difference in *variances*. Differences in variances affect the denominator in the calculation of the t-value, so it might be important to know this information, as you will see. Now you have to decide the main question as to whether there is a difference in *means*. Note that there are two t-values calculated: The one following the phrase "equal variances assumed" is –10.148, and the one following "equal variances not assumed" is –9.764. Because variances are assumed equal, you follow the t-value in the first line and see that its significance level is less than .05. In fact, it is so small, it's below .000; you only see the first few digits, and the actual number could be something like .00059, for example. You can now say that the probability of obtaining a t-value of 10.148, with samples of this size and with these standard deviations, by chance alone is less than 1 in 1,000 ($p < .001$).

This is certainly good enough for you to tweet your results to the world and state that indeed men are taller than women in your sample. The difference between means is –5.33 inches. The minus signs are simply a function of which mean was subtracted first. Because the men's mean was subtracted from the women's, it results in a minus sign. If women's mean height were subtracted from the men's, then the result would still be 5.33 inches, but positive. The number is important for determining the significance level, not the minus or plus sign.

Confidence Interval. If you were to draw 100 samples of the size of the sample in Table 8.1, 95 of them would have a mean difference in heights somewhere between 4.29 and 6.37 inches; this is called the *confidence interval* and is reported in the column "95% Confidence Interval of the Mean." In other words, you are 95 percent confident that the true difference in mean heights between men and women in the population from which you drew this particular sample is within that range. Remember you are inferring population information based on what you found in one particular sample, so there is likely some error and a range within which the true information lies. And it also assumes you are drawing a random probability sample from that population.

BOX 8.2
T-TESTS IN ACADEMIC ARTICLES

Simon and Furman (2010) studied the role of high school seniors' perception of their parents' conflicts and the impact on their own romantic relationships. Here are some of the results, looking at gender differences. High scores on "interparental conflict" reflect perception of many parental arguments, high scores on "relationship conflict" in students' romantic relationships indicate much antagonism, and higher scores on "strategies" reflect physical aggression (like pushing, shoving) and conflict engagement ("throwing insults and digs").

MEASURE	BOYS' MEAN (SD)	GIRLS' MEAN (SD)	T-VALUE
Interparental conflict	3.20 (1.31)	3.74 (1.38)	−1.34*
Romantic relationship conflict	2.43 (1.01)	2.43 (1.01)	0.42
Romantic conflict strategies			
Aggression	1.48 (0.79)	1.52 (0.68)	0.26
Engagement	2.28 (1.09)	2.80 (1.26)	−2.49*

* $p < .05$

Comparing the mean scores between the high school boys and girls, two independent groups, calls for a t-test. The results presented here show that girls are significantly more likely to perceive parental conflict and more likely to use conflict engagement as a romantic relationship strategy, but neither boys nor girls are more likely to have romantic relationship conflicts or to use aggression as a technique.

Notice that it usually takes a t-value closer to 2 to be statistically significant at the .05 level. It's easy to remember: It usually takes "t for two" to be significant!

Paired Samples T-Test

The previous analysis was based on the assumption that the two categories being compared do not have any overlapping membership. These are independent samples. Sometimes, however, you need to compare two means for the same sample of people. For example, you might want to compare graduating seniors' average grades with their own first-year GPAs. Or you might want to look at the before and after effects of a new diet plan and compare the subjects' weights when they first started the diet with their current weights. When you do this, you must also take into account that there is a connection or correlation between the two measures; after all, it is the same pool of people being studied at two different points in time. You need to use a t-test formula that assesses the difference in means while taking into account the correlation between the two measures. In such cases, we calculate a *paired samples t-test*. Consider the following two measures: starting salary and current salary (Table 8.2).

The same 474 employees were sampled at two different periods of their employment. The average beginning salary was $17,016.09 and the average current salary is $34,419.57. Certainly there is a difference, but is it significant? Or should the employees complain more about their low salary increases using the null hypothesis that there is no significant difference between their current salary and what they started out with? Remember this is not about any one person's change in salary, but change in the *average* salary of all these 474 employees together.

According to the results of the SPSS analysis, the difference in average salaries of $17,403.48 is statistically significant. The probability of obtaining a t-value of 35.036 by chance, with a standard error of the mean of 496.73 and 474 subjects, is less than .001. And we are 95 percent confident that the true difference between average beginning and current salaries in the population from which this sample is drawn is somewhere between $16,427.41 and $18,379.56.

One-Sample T-Test

There are also a few occasions when we might want to compare a mean to some standard that already is known. For example, you wonder if the average GPA of a group of students in each major is really different from the average GPA of all students at the university. In this case, you would calculate a *one-sample t-test* (Table 8.3). Let's say you find out the grade point average for students in one particular major is 3.10 (out of 4.0). According to the registrar's office, the average GPA for the entire college is 2.95, so this number is entered as the one to compare with the average for your sample. But is that difference of 0.1513 GPA units statistically significant? It looks so small; it's difficult to tell if it is a meaningful difference unless you perform a test of statistical significance. Using the one-sample t-test, you learn that the probability of obtaining a t-value of 4.44 by chance is less than .001 and is therefore statistically significant. You can conclude that the sample of 150 has a larger mean GPA than the university as a whole. There's something special about this group—they must be sociology students!

ANALYSIS OF VARIANCE (ANOVA)

With many research questions and hypotheses, we find ourselves wanting to compare more than two means at a time. For example, we might be interested in studying racial/ethnic differences in the amount of television viewing. When there are more

Table 8.2 Example of Paired Samples T-Test, SPSS

PAIRED SAMPLES STATISTICS

	Mean	N	Standard Deviation	Standard Error Mean
Pair 1 Current salary	34,419.57	474	17,075.66	784.31
Beginning salary	17,016.09	474	7,870.64	361.51

PAIRED SAMPLES CORRELATIONS

	N	Correlation	Significance
Pair 1 Current salary and beginning salary	474	0.880	.000

PAIRED SAMPLES TEST

PAIRED DIFFERENCES

	Mean	Standard Deviation	Standard Error Mean	95% CONFIDENCE INTERVAL OF THE DIFFERENCE Lower	Upper	t	df	Significance (2-tailed)
Pair 1 Current salary minus beginning salary	17,403.48	10,814.62	496.73	16,427.41	18,379.56	35.036	473	.000

Table 8.3 Example of One-Sample T-Test, SPSS

			ONE-SAMPLE STATISTICS	
	N	Mean	Standard Deviation	Standard Error Mean
GPA	150	3.1013	0.4171	3.4E–02

ONE-SAMPLE TEST

TEST VALUE = 2.95

					95% CONFIDENCE INTERVAL OF THE DIFFERENCE	
	t	df	Significance (2-tailed)	Mean Difference	Lower	Upper
GPA	4.444	149	.000	0.1513	8.4E–02	0.2186

than two categories for race/ethnicity, which is typically the case, we should use analysis of variance (ANOVA), a technique that asks whether the differences *within* a category are larger or smaller than those *between* (or, more grammatically, *among*) three or more categories. Like the t-test, ANOVA requires that the dependent variable be interval/ratio, a dichotomy, or an ordinal measure with equal-appearing intervals, so that a mean and variance can be calculated. The categories of the independent variable (called a *factor* in ANOVA) should be mutually exclusive in order to perform a one-way ANOVA. When there are multiple factors or independent variables, and interactive effects among those factors need to be assessed, or there are repeated measures on the same sample to be compared, use more sophisticated versions of ANOVA as discussed in advanced statistics books and on online websites.

Understanding ANOVA

ANOVA works by looking at the dispersion (variation) of scores around several means: the overall total dependent variable mean and the individual means for each of the categories of the main factor or independent variable. The computer program calculates an F-value, which, if significantly high enough, suggests that there is indeed a difference among the three or more means. In short, it tells you that the independent variable or factor has some effect on the dependent or outcome variable.

Consider the example in Table 8.4: Your hunch is that students in different years of college have different grade point averages. Everyone likes to think that their class is smarter than the others! Because grade point average is an interval/ratio measure, you calculate a mean and standard deviation for students in each level of college. As can be seen, there is some difference, but you need to be sure that the difference is not one that could have occurred merely by chance alone more than five times out of every 100 samples ($p > .05$). If it is, then those differences are not statistically significant, and we accept the null hypothesis of no difference.

The first part of the ANOVA output is descriptive data that show that seniors have higher GPAs (3.14) than others and higher than the overall average for all 150 students (3.10). Staying in college must make you smarter. But you clearly can't make any causal conclusions until you can first demonstrate that there is a relationship and rule out alternative explanations (those with lower grades were kicked out). Since you are comparing means for more than two categories of the main factor (four levels of the independent variable are measured), you use ANOVA, which calculates an F ratio comparing the variation between the four categories relative to the variation within categories.

Table 8.4   Example of Analysis of Variance, SPSS

DESCRIPTIVES

GPA	Mean	Standard Deviation
Freshmen	3.0714	0.53763
Sophomores	2.9130	0.46251
Juniors	3.1361	0.40867
Seniors	3.1424	0.38201
Total	3.1013	0.41708

ANOVA

GPA	Sum of Squares	df	Mean Square	F	Significance
Between groups	0.995	3	0.332	1.942	.125
Within groups	24.925	146	0.171		
Total	25.920	149			

According to the second part of the table, the F-value of 1.942 is not significant at .125, which is greater than the minimum .05 we require for statistical significance. Therefore, we accept the null hypothesis of no difference and conclude that students' average grades cannot be attributed simply to what year of college they are in. One class does not appear to be smarter than the others.

As a rule of thumb, if the between groups variation is the same as the within groups variation (a null hypothesis of no difference), then the F statistic will be 1, since dividing a number by itself always results in one. In this situation, there is no influence of the independent variable categories on the dependent variable. The larger the ratio is, the more likely it will be statistically significant. However, it is not a measure of strength like a Pearson r correlation, so the size of F does not tell you much. More variation among the three or more groups in comparison to the variation within the groups results in an F increasingly larger than 1 and likely to be statistically significant. In other words, the independent variable categories make a difference in understanding the dependent variable when the between groups variation is larger than the within groups variation.

Like the Levene F-test for equality of variance used in the independent samples t-test, ANOVA involves analyzing (the "AN" of ANOVA) whether the two variances (the between and within ones, the "VA" of ANOVA) are equal or not. If the probability of obtaining that F by chance is less than .05, we conclude that there is a significant difference in variances; we reject the null of no difference and agree that the between groups variance is larger. We then state that this is due to at least two of the categories having very different means from each other, because the original formula calculated how much these means differed from the overall mean. And that is why something called an analysis of *variance* is used to understand the difference among three or more *means*!

DIFFERENCES NOT STRENGTH

With both the t-test and the F-test, we learn about the differences between means among several subgroups or samples. These are alternative ways of providing information about the relationship between two variables. They do not tell us, like Pearson r, what the strength of the relationship is, or how much variance is explained in the dependent variable by the independent as PRE statistics do. Many researchers make use of several statistical techniques at the same time to gather as much information as they need about their hypothesized relationships. It is not unusual then to see a study employing crosstabs, correlation coefficients, and differences in means tests for the same research questions and hypotheses. Each provides alternative insights about the relationships. Your job as a researcher is to make the decisions

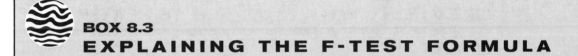

BOX 8.3
EXPLAINING THE F-TEST FORMULA

The F is based on a formula that breaks down *the total sum of squares* (which is the sum of squaring each score subtracted from the overall mean) into the *within groups sum of squares* and the *between groups sum of squares*. Within groups sum of squares is calculated by taking each score, subtracting its category (or subsample) mean from it, squaring the difference, and summing those numbers:

$$\sum_{i=1}^{k}\left[\sum_{j=1}^{n_i}\left(X_{ij} - \overline{X}_i\right)^2\right]$$

Or it can be calculated by subtracting the between sum of squares from the total sum of squares:

$$SS_w = SS_{tot} - SS_B$$

The between groups sum involves subtracting the overall mean from each category or group mean, squaring the difference, and summing those numbers:

$$SS_B = \sum N_i \left(\overline{X}_i - \overline{X}_{tot}\right)^2$$

The total sum of squares formula is

$$SS_{tot} = \sum X_{tot}^2 - \left(\frac{(\sum X_{tot})^2}{N}\right)$$

The F-value is the ratio of the between groups variation (the between group sum of squares divided by the *df*, or what is called the *mean squares*) to the within groups variation (the within groups sum of squares divided by the *df*):

$$F = \frac{ms_{Between}}{ms_{Within}}$$

Using the GPA example in Table 8.4, the F is not statistically significant because the variation (between groups mean squares of 0.332, which comes from dividing the between sum of squares, 0.995, by 3, the *df*) of the category means (3.07 for first-year students, 2.91 for sophomores, etc.) from the overall mean (3.10) is not larger than the variation (within groups) of the first-year students' GPAs from their category mean of 3.07, plus the variation of the sophomores' GPAs from their group's mean of 2.91, and so on.

about which statistics best assist you in answering your questions and learning what the statistics differently contribute to your understanding of the relationships.

We turn our attention in the next chapter to what happens when we want to analyze three or more variables at the same time. Social reality is such that behavior and attitudes are rarely explained by simple bivariate relationships. We often need to look at more complex relationships among multiple variables.

REVIEW: WHAT DO THESE KEY TERMS MEAN?

ANOVA (F Value)
Between groups and
   within groups
   variances
Confidence interval

Independent samples t-test
Levene's F for equality
   of variance
One-sample t-test

Paired samples t-test
Standard error of the
   difference between means
T-Tests

TEST YOURSELF

1. What *statistic* would be most useful to test whether you reject or accept each of the following hypotheses? (Choose the *best* one if more than one can be used.)
   a. There is a difference in GPA between students living in Mead Hall, Atherton Hall, and Holden Hall.
   b. There is no difference in employees' mean scores on a scale (from 1 = terrible to 10 = wonderful) used to measure opinions about the food in the company's dining hall. The survey was given both before and several months after a new company was hired to prepare the food.
   c. There is no difference in average number of books read in the past year between Democrats and Republicans.
2. The General Social Survey measured attitudes toward classical music (where 1 = like very much and 5 = dislike very much) and the differences between those who graduated from college and those who didn't (see Table 8.5).
   a. Why is a t-test being used here?
   b. What does the Levene test tell you?
   c. Which t-value do you use then?
   d. Interpret the results for those who understand statistics.
   e. How do you put the results into words for the general public?

INTERPRET: WHAT DO THESE REAL EXAMPLES TELL US?

1. Meyer et al. (2010) report on how credibility is viewed by people using the Internet for the news. Although they studied various types of websites (blogs, straight news, etc.) on a variety of perceptions (expertise, opinionated, credible, etc.), let's take a look at just a few of the results.

   "Expertise" meant reporters sounded like experts, knew what they were talking about, and had done their homework for various types of Internet news sites.

Table 8.5 Example of General Social Survey

GROUP STATISTICS

			N	Mean	Standard Deviation	Standard Error Mean
Classical music	College degree	No college degree	1,080	2.85	1.224	0.037
		College degree	344	2.06	0.983	0.053

INDEPENDENT SAMPLES TEST

		LEVENE'S TEST FOR EQUALITY OF VARIANCES		T-TEST FOR EQUALITY OF MEANS		
		F	Significance	t	df	Significance (2-tailed)
Classical music	Equal variances assumed	47.774	.000	10.913	1,422	.000
	Equal variances not assumed			12.205	709.993	.000

A "straight news" site represented standard objective news writing, a "blog" was a conversational format of personal views, a "collaborative" was cowritten by a reporter and local citizens, and an "opinionated" site was strong in its opinions, like Fox News.

Each of the 140 college students read four different stories and was asked to rate his or her perception of expertise. Higher means indicate stronger belief in the expertise level of the news site. Paired t-tests were calculated between "straight news" and each of the other three website formats.

Expertise	Mean	T-Value
Straight news	3.53	
Opinionated	2.81	6.85**
Collaborative	3.24	3.39**
Blog	2.82	7.45**

$df = 106$, ** $p < .01$

a. Why is a paired t-test used for these data?
b. State the null hypothesis for this study.

c. Explain to someone knowledgeable about statistics what is going on with the t-values and significance levels. And put into words for a general audience the conclusions you would draw from these results.

2. Here is some SPSS output from the General Social Survey assessing whether people with different outlooks on life (exciting, routine, dull) have different years of education on average. In short, does education make a difference in how people feel about life?

DESCRIPTIVES

HIGHEST YEAR OF SCHOOL COMPLETED

	N	Mean	Standard Deviation	Standard Error	95% Confidence Interval for Mean	
					Lower Bound	Upper Bound
Exciting	433	13.64	3.108	0.149	13.35	13.94
Routine	503	12.44	2.726	0.122	12.20	12.68
Dull	41	10.49	3.257	0.509	9.46	11.52
Total	977	12.89	3.022	0.097	12.70	13.08

ANOVA

HIGHEST YEAR OF SCHOOL COMPLETED

	Sum of Squares	df	Mean Square	F	Significance
Between groups	582.721	2	291.360	34.077	.000
Within groups	8,327.779	974	8.550		
Total	8,910.499	976			

a. Why is an F-test (ANOVA) the best statistic to use here?
b. State a two-tailed null hypothesis that these statistics test.
c. Interpret the results in words. What do you conclude (1) for the statistical experts, and (2) for members of the general public who don't understand these numbers?

CONSULT: WHAT COULD BE DONE?

1. You've been asked by the university's financial aid office to compare the cost of textbooks used in five different majors: sociology, psychology, biology, history, and women's studies. Explain the steps you would take to do this and what statistic you would use to help you make a conclusion.
2. You've been asked by the human resources office at a company to assist in resolving a rumor going around the workplace. It seems that people believe that higher educated employees are getting larger salaries than less educated ones. Describe what research you would do and the statistical procedures you would use to gather information to address the rumor.

DECIDE: WHAT DO YOU DO NEXT?

For your study on how people develop and maintain friendships, as well as the differences and similarities among diverse people, respond to the following items:

1. Look over your hypotheses and list the ones that could be assessed using a t-test. If none, develop a research question that uses t-tests. What would the independent variable look like, and what information would you need for the dependent variable?
2. List the hypotheses that could be assessed using ANOVA. If none, develop a research question that uses ANOVA. What would the independent variable look like, and what information would you need for the dependent variable?
3. If you have actual data from a survey, develop a few hypotheses suitable to analyze with at least one t-test and one ANOVA. Interpret the results and put into words what you have discovered.

9 ANALYZING DATA

Multiple Variables

Nothing in life is to be feared, it is only to be understood. Now is the time to understand more, so that we may fear less.

—*Marie Curie, Nobel laureate scientist*

LEARNING GOALS

This chapter focuses on the analysis of three or more variables to answer more complex research questions. It discusses when to use various kinds of multivariate analyses and how to elaborate your findings with additional variables. By the end of the chapter, you should be able to interpret multiple regression analyses and perform elaboration techniques with control variables.

Figure 9.1 Statistical Decision Steps

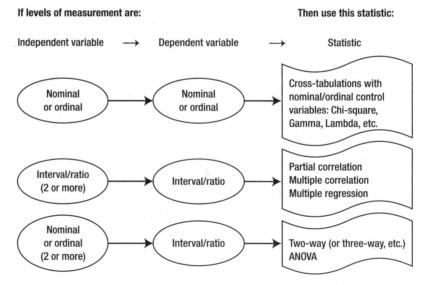

Although it may be tempting to try to understand people's behavior and opinions with just one simple explanation or cause, in reality there are multiple reasons for those beliefs and actions. Understanding more about how your variables work together in explaining behavior and attitudes is essential for doing quality research. If, for example, you are trying to understand why students' grades vary, the number of hours studying is probably just one reasonable answer. What else might explain differences in grades? Quality of the instructor, test-taking abilities, physical and mental health of the people at the time of doing the required work or test, difficulty of an assignment, basic intelligence levels, and the party the night before are all plausible alternative explanations.

One of the goals of research is to evaluate multiple explanations or predictors of some dependent variable or outcome. Sometimes it is also necessary to verify that the relationships we have uncovered withstand alternative explanations or are consistent for a variety of subgroups in the sample. Unless the elimination of other variables that could possibly explain the main relationship occurs, we have not established cause and effect. Recall the three elements discussed in Chapter 1 that are essential to declaring causality. Most advanced methods books and online websites describe multivariate statistics in greater depth, but for our introductory purposes, this chapter focuses on some basic techniques and concepts that are helpful for analyzing data for the first time

ELABORATING RELATIONSHIPS: CONTROL VARIABLES

A typical question is whether the relationships you have discovered are strong enough to withstand other plausible explanations. What we need to do is something called *elaboration,* which extends our knowledge about the association to see if it continues or changes under different situations (see Rosenberg 1968). Researchers who use experimental designs use a control group as a means of verifying the results that occurred in the experimental group. Similarly, those of us doing survey research can control for other variables during the data analysis phase.

Consider an association that is statistically significant between respondents' educational level and yearly family income (Table 9.1). Let's first take the case where these variables are ordinal measurements—education: less than high school, high school graduate, and so on; income: under $25,000, $25,000 to $39,999, and so forth. Minimally, a chi-square can be calculated for this cross-tabulation of data, along with some other statistics appropriate for ordinal measures, such as Gamma or Kendall's Tau-c. Perhaps you know, though, that women often earn less than men, and you wonder if educational level remains as good a predictor of total family income for women. The question is whether the relationship is still statistically significant when *controlling* for sex of the respondent; that is, does it still apply equally to men and women? Sex becomes the test or control variable used to elaborate the original relationship between education and income.

Table 9.1 Cross-Tabulation of Income and Education, SPSS

TOTAL FAMILY INCOME BY HIGHEST DEGREE CROSS-TABULATION

			HIGHEST DEGREE					
			Less Than High School	High School	Junior College	Bachelor	Graduate	Total
Total family income ($)	24,999 or less	Count % of highest degree	196 70.3%	315 40.4%	25 27.8%	39 16.7%	9 8.0%	584 39.0%
	25,000 to 39,999	Count % of highest degree	28 10.0%	175 22.4%	21 23.3%	58 24.8%	18 15.9%	300 20.1%
	40,000 to 59,999	Count % of highest degree	16 5.7%	121 15.5%	23 25.6%	52 22.2%	18 15.9%	230 15.4%
	60,000 or more	Count % of highest degree	39 14.0%	169 21.7%	21 23.3%	85 36.3%	68 60.2%	382 25.5%
Total		Count % of highest degree	279 100.0%	780 100.0%	90 100.0%	234 100.0%	113 100.0%	1,496 100.0%

CHI-SQUARE TESTS

	Value	df	Asymptotic Significance (2-tailed)
Pearson chi-square	264.299[a]	12	.000
N of valid cases	1,496		

[a] Zero cells (0.0%) have expected count less than 5. The minimum expected count is 13.84.

SYMMETRIC MEASURES

		Value	Approximate Significance
Ordinal measures	Kendall's Tau-c	0.295	.000
	Gamma	0.457	.000
N of valid cases		1,496	

As the chi-square and Gamma indicate, there is a significant relationship, and those with less education tend to be in households with less family income: 70.3 percent of those with less than a high school degree earn under $25,000 yearly, compared with 60.2 percent of those with a graduate degree who earn over $60,000 a year. Now let's see how well this relationship holds up when introducing a third variable: respondents' sex, which has two categories or values, male and female (Table 9.2).

Partialling

Two tables are developed, one showing the relationship between education and family income for males and one for females, along with two sets of chi-squares and Gammas. There are now three variables involved in the multivariate analysis. When you introduce a control variable, it is referred to as first-order *partialling*. You can continue to add multiple variables, called second-order, third-order, and so on, for more elaborate models, but interpretation can get complex at that point, especially if each of those variables has numerous values or categories. The original bivariate association is the *zero-order* relationship.

Notice in the new partial tables that all the statistics remain significant, and the ordinal measures of strength (Gamma and Kendall's Tau-c) are about the same, although slightly stronger for men. You have replicated your original finding and can now conclude that there is a strong relationship between education and income, because it holds up even when controlling for sex. It is unlikely that being male or female directly affects total family income level; education appears more important. For example, 65.6 percent of men with less than a high school education tend to have under $25,000 a year in total family income; 74 percent of women with less than a high school education earn under $25,000 a year. This pattern is replicated across the other education and income categories. You now might want to test the original zero-order relationship again with a new control variable to see if it is sustained for other conditions.

Spurious and Antecedent Relationships

What would it mean when you introduced a third control variable and the relationship did not hold up? If the original relationship disappears or becomes less strong—it is statistically significant for neither the men nor the women, or the correlation coefficients decline substantially—you would claim the original relationship was *spurious*. You would verify this by creating a cross-tabulation for sex and education and another one for sex and income and see if indeed sex was *antecedent* to or

Table 9.2   Cross-Tabulation of Income and Education, Controlling for Sex, SPSS

**TOTAL FAMILY INCOME BY HIGHEST DEGREE
BY RESPONDENT'S SEX CROSS-TABULATION**

Respondent's Sex				Less Than HS	High School	Junior College	Bachelor	Graduate	Total
				HIGHEST DEGREE					
Male	Total family income	24,999 or less	Count % of highest degree	82 65.6%	109 35.9%	11 29.7%	17 15.7%	5 7.5%	224 34.9%
		25,000 to 39,999	Count % of highest degree	22 17.6%	78 25.7%	6 16.2%	26 24.1%	11 16.4%	143 22.3%
		40,000 to 59,999	Count % of highest degree	9 7.2%	62 20.4%	10 27.0%	27 25.0%	10 14.9%	118 18.4%
		60,000 or more	Count % of highest degree	12 9.6%	55 18.1%	10 27.0%	38 35.2%	41 61.2%	156 24.3%
	Total		Count % of highest degree	125 100.0%	304 100.0%	37 100.0%	108 100.0%	67 100.0%	641 100.0%
Female	Total family income	24,999 or less	Count % of highest degree	114 74.0%	206 43.3%	14 26.4%	22 17.5%	4 8.7%	360 42.1%
		25,000 to 39,999	Count % of highest degree	6 3.9%	97 20.4%	15 28.3%	32 25.4%	7 15.2%	157 18.4%
		40,000 to 59,999	Count % of highest degree	7 4.5%	59 12.4%	13 24.5%	25 19.8%	8 17.4%	112 13.1%
		60,000 or more	Count % of highest degree	27 17.5%	114 23.9%	11 20.8%	47 37.3%	27 58.7%	226 26.4%
	Total		Count % of highest degree	154 100.0%	476 100.0%	53 100.0%	126 100.0%	46 100.0%	855 100.0%

CHI-SQUARE TESTS

Respondent's Sex		Value	df	Asymptotic Significance (2-tailed)
Male	Pearson chi-square	136.855	12	.000
	N of valid cases	641		
Female	Pearson chi-square	145.194	12	.000
	N of valid cases	855		

SYMMETRIC MEASURES

Respondent's Sex			Value	Approximate Significance
Male	Ordinal measures	Kendall's Tau-c	0.341	.000
		Gamma	0.491	.000
	N of valid cases		641	
Female	Ordinal measures	Kendall's Tau-c	0.261	.000
		Gamma	0.433	.000
	N of valid cases		855	

came before the dependent variable of income, and if sex explained the dependent variable of education. If so, then the original relationship had the illusion of an association only because the independent and dependent variables were both related to a third antecedent variable for which you controlled.

Here is a classic example of a spurious relationship: The number of fire engines that show up at a fire (independent variable) is significantly related to the amount of the fire damage (dependent variable). In other words, you might conclude that if only one or two fire trucks appeared, the damage would have been less. Here you can see the problem of making a cause and effect conclusion based on just a significant correlation. Clearly, there is need for an alternative explanation, and perhaps size of the fire is one such control or test variable. Sure enough, the original relationship disappears when you demonstrate that fire size (independent variable) is highly correlated to the number of fire trucks that are called (dependent) and that fire size (independent) is highly related to the amount of damage (dependent). Fire size is the more direct reason for the amount of damage, not the number of fire engines!

Specification

Sometimes when you control for a third variable, you notice that the original relationship disappears only for some of the values of the test variable. Let's say that an elaboration of the education and income model showed that the original relationship applied only to the men in the sample and not to the women. You would notice this by seeing that the statistical tests remained significant for the men but were no longer statistically significant for the women or they were weakened. In such cases, you have *specification* because you determined very specific situations in which the relationship holds and those in which it does not.

Suppressing Relationships

Elaboration is also useful when you find no relationship or a very weak one between your independent and dependent variables. Don't give up and think that you found little. Elaborate on this nonfinding and introduce a control variable to see if for some reason this is *suppressing* or holding back the original zero-order relationship. Imagine that we found a weak relationship between parents' educational level and their adult children's (the respondents') income, despite reports about greater income for those growing up in higher educated families. So you introduce respondents' education as a test variable and see whether the weak relationship between parents' education and income continues for each level of respondents' education. If a relationship now appears in one or more of the respondents' education categories, we can conclude that education was suppressing the original relationship.

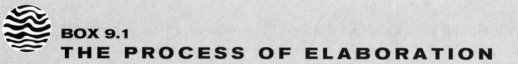

BOX 9.1
THE PROCESS OF ELABORATION

Zero-order relationship: X (independent variable) → Y (dependent variable)
First-order partialling: Z (control/test variable with three values or categories)

1. Antecedent   Z → X → Y original relationship elaborated

2. Intervening   X → Z → Y relationship works through the test variable

3. Spurious   X → Y relationship is significant
                  for Z (1), X → Y relationship disappears/weakens
                  for Z (2), X → Y relationship disappears/weakens
                  for Z (3), X → Y relationship disappears/weakens
                  because of antecedent Z → X
                                        Z → Y

BOX 9.1 CONTINUED

4. Specification   for Z (1), X → Y relationship disappears/weakens

for Z (2), X → Y relationship remains (replicated)

for Z (3), X → Y relationship remains (replicated)

or any other combination where the relationship disappears, remains the same (replicated), or is reversed, in some but not all of the categories of the test variable

5. Suppressor   X → Y relationship is weak or none

for Z (1), X → Y relationship appears

for Z (2), X → Y relationship appears

for Z (3), X → Y relationship appears because the control variable weakened or suppressed the original relationship.

Intervening Variables

Now imagine that you find that a strong relationship occurs within each level of respondents' education. In other words, for those with less than high school education, there is now a strong relationship between parents' education and income; among those who are high school graduates, there is also a strong relationship between parents' education and income; and so on. What you have shown is that respondents' education was suppressing the original relationship because it intervenes between parents' educational level and income. Respondents' education is an *intervening* variable that was mediating and keeping the original relationship weak, since parents' education is only an indirect influence on respondents' income. Parents' education is highly correlated with respondents' education, which in turn predicts income.

Partial Correlations

When variables are interval/ratio measures, controlling for a test variable can be done using partial Pearson r correlation (partial Tau and partial Gamma are also available for ordinal measures). A *partial correlation* takes into account the relationship between X and Y when the control variable Z is introduced. It calculates the Pearson r correlation between X and Y, X and Z, and Y and Z and then recalculates the original zero-order relationship between X and Y after the effect of Z has been partially removed. For example, a Pearson r correlation of 0.084 was calculated for seniority on the job and current salary. That very low coefficient seemed to indicate that there was no statistically significant relationship between these two variables, somewhat counterintuitive to what you might think. After all, the longer people work someplace, the more raises they get and their salaries increase.

However, it's also possible that those who worked at the company the longest started at lower salaries than those who began more recently, given the realities of inflation. Starting salary might be an antecedent variable that could elaborate on the original relationship found. So, a partial correlation was calculated, removing the relationship between starting salary and current salary, and sure enough the original relationship increases to 0.214 (Table 9.3). Starting salary (the control variable Z) was suppressing the original relationship between seniority and salary. Even though it is statistically significant, it remains somewhat low in strength, suggesting that there are other reasons besides seniority to explain current salary.

There are numerous ways to introduce additional variables for elaboration and to test various scenarios. For example, some categories of a political party control variable (such as "Independents") may be suppressing the original relationship, while other categories do not (such as "Democrats" and "Republicans"). Or an original relationship might be a strong positive one, but after controlling for a third variable, it becomes a strong negative one in some categories of the control variable but remains positive in the others. The important point to remember is that in order to establish any kind of cause and effect relationship you must be able to rule out alternative plausible explanations; the process of elaboration is a good way of doing this and a good way of uncovering other kinds of relationships among your variables.

Table 9.3 Example of Partial Correlation, SPSS

	CORRELATIONS	
		Seniority Months since Hire
Current Salary	Pearson Correlation	0.084
	Significance (2-tailed)	.067
	N	474

Partial Correlation Coefficients

Controlling for ... Beginning salary
seniority

Current salary	0.2138
	$p = .000$

MULTIPLE RELATIONSHIPS

Using control variables is one kind of multivariate analysis, primarily focused on providing more elaborate understanding of initial relationships. Another important type is an analysis of additional independent variables that might contribute to a better and more complete understanding of the dependent variable or outcome. When these variables are interval/ratio measures or ordinal measures with equal-appearing intervals—instead of the nominal and ordinal ones better suited for cross-tabulations and the elaboration analyses just described—then some fairly sophisticated statistical procedures can be used.

Two-Way ANOVA

Let's say you are interested in the differences in mean income among several racial/ethnic groups. As you recall, for comparing means among three or more nominal or ordinal categories, use one-way analysis of variance (ANOVA). If the probability of obtaining the F-value is less than .05, you declare that a relationship exists between race/ethnicity and income and specify which groups obtain higher average salaries. Yet, this seems too simple an explanation, so you introduce a third variable.

You could control for sex by rerunning the ANOVA for men and then another one for women, but another technique would be to introduce sex as a third variable (sometimes called a factor or covariate) and perform a *two-way analysis of variance,* or three-way or more, depending on how many new factors are included. In addition, you might want to see if there is another factor affecting the outcome that is a result of the possible combined impact of the two (or more) independent variables, not just their individual effects. This is what's called an *interactive effect.* For example, race might affect income, and sex might also have a main effect on income, but the combined impact of race and sex might be even stronger. The computer calculates the F-values for the main effects of each of the independent variables, in this example sex and race, along with the F-value of any interactive effects, sex by race. It is interpretable similarly to the F calculated in a one-way ANOVA. If F is statistically significant, the ANOVA would suggest that, for example, women of color receive lower salaries than white women, and so on. Details for doing multivariate and factorial ANOVA are available in advanced statistics books and online.

Multiple R

Another very common multivariate technique is *multiple correlation.* This is based on the Pearson r correlation coefficient and essentially looks at the combined effects of two or more independent variables on the dependent variable. These variables should be interval/ratio measures, dichotomies, or ordinal measures with equal-appearing intervals, and assume a linear relationship between the independent and dependent variables. Rather than partialling out the impact of a third variable to see what remains of the original relationship, multiple correlation includes the effects of a third or fourth, or more, variable on the dependent variable. It is represented as the capital letter R and, similar to r, is a PRE (proportional reduction in error) statistic when squared. However, unlike bivariate r, multiple R cannot be negative because it represents the combined impact of two or more independent variables, so direction is not given by the coefficient.

What multiple R^2 tells us is the proportion of the variation in the dependent variable that can be explained by the combined effect of the independent variables. Let's say you found that current salary depends on three explanations: starting salary, seniority, and age of employee. Each of these independent variables is correlated with current salary, but you want to know their combined impact. The result is not simply the addition of the bivariate Pearson r correlations for each of these variables and the dependent one, because some variables, like age and seniority, for example, are correlated with one another, and this overlap needs to be taken into account mathematically.

The computer calculates a very strong positive multiple R correlation coefficient of 0.897 and an R^2 of 0.805. This tells you that approximately 80.5 percent of the variation in current salaries in your sample is accounted for by a combination of starting salary, seniority, and age. Which ones contribute more or less cannot be determined just by looking at the multiple correlation; however, as discussed in the next section, there is a way of getting this information. R^2 also tells you that if you wanted to predict salaries, you would reduce your errors in predicting those figures by around 80 percent knowing three things: what the starting salaries were, how long the employees worked for the company, and ages of the workers.

Linear Multiple Regression

Uncovering which independent variables are contributing more or less to the explanations and predictions of the dependent variable is accomplished by a widely used technique called *linear regression*. It is based on the idea of a straight line that has the formula $Y = a + bX$, where

- Y is the value of the predicted dependent variable, sometimes called the criterion and in some formulas represented as Y' to indicate Y-predicted;
- X is the value of the independent variable or predictor;
- a is the constant or the value of Y when X is unknown, that is, zero; it is the point on the Y axis where the line crosses when X is zero; and

- b is the slope or angle of the line and, because not all the independent variables are contributing equally to explaining the dependent variable, b represents the unstandardized weight by which you adjust the value of X. For each unit of X, Y is predicted to go up or down by the amount of b.

The Regression Line. Pearson r correlations and regressions assume linearity. If you were to construct a scatterplot, and the correlation was a perfect 1.0, then all the points representing a score on X and a score on Y would fall into a straight line. Most of the time, of course, a correlation is not perfect, so the points are more likely not to form a straight line. Therefore, through the scatterplot, a best-fitting line is drawn, around which the points are the closest. This is similar to the idea of a mean being the point around which the scores in a distribution are the closest. This best-fitting line is called the *regression line*, which predicts the values of Y, the outcome variable, when you know the values of X, the independent variables. Linear regression analysis calculates the constant (a), the coefficient weights for each independent variable (b), and the overall multiple correlation (R). Preferably, low intercorrelations exist among the independent variables in order to find out the unique impact of each of the predictors. This is the formula for a multiple regression line:

$$Y' = a + bX_1 + bX_2 + bX_3 + bX_4 \ldots + bX_n$$

Methods for Entering Variables. One method for generating a regression equation is to enter the independent variables in the order you believe—based on theory or prior research—they contribute to explaining the dependent variable. Another is to use what is termed *stepwise* multiple regression, which lets the computer enter the variables, either individually or in blocks or groups of variables, in the order of their correlation to the dependent variable. It selects the strongest predictor first, then the next, and so on, one at a time from the entire list of independent variables (or from within each block) until a variable is not statistically significant to enter the equation, usually set at the .05 level of significance. Another rule of thumb is to be sure that each new variable entered contributes at least 1 percent to the overall explanation of the variance of the dependent variable.

Beta Coefficients. The result is an equation that can be used to predict values of the dependent variable. The information provided in the regression analysis includes the b coefficients for each of the independent variables and the overall multiple R correlation and its corresponding R^2. Assuming the variables are measured using different units as they typically are (such as pounds of weight, inches of height, or scores on a test), then the b weights are transformed into standardized units for comparison purposes. These are called Beta (β) coefficients or weights and essentially are interpreted like correlation coefficients: Those farthest away from zero are the strongest, and the plus or minus sign indicates direction. If all the variables are measured in the same units, or if you are comparing a particular variable in one regression equation with the same one in another equation, then the unstandardized b coefficients are easily interpreted. With the Beta coefficients and the R^2, you have the most relevant information you need to arrive at some conclusions.

A multiple regression provides a description of possible influences on the dependent variable. It is a statistical technique designed especially for explanatory and predictive research hypotheses. Multiple regression is a robust statistical procedure that allows some flexibility in the kinds of variables used. Ideally, your measures are interval/ratio ones that have a linear relationship between the independent and dependent variables. If your measures are not interval/ratio, then determine whether any ordinal measures have equal-appearing intervals (such as Likert-type scales, or if income is in categories with equal ranges) or there are dichotomous variables with two categories.

However, you cannot use nominal measures with three or more categories. Typically, you have to create dichotomous or *dummy variables* for nominal variables, such as religion. Because there is no order to the religions listed in your questionnaire, you might recode your data into Religious, Not Religious; or Protestant, Not Protestant; or Christian, Not Christian; or Muslim, Not Muslim; or any other possible dichotomy, depending on the goals of your research. The same applies to such nominal demographic variables as race/ethnicity, sexual orientation, political party affiliation, and other variables with several nominal categories.

Let's hypothesize that there is a relationship between students' grades in college (CUM GPA) and their high school grades (HS GPA), total scores on a standardized test (TOTSAT), and their participation in extracurricular activities (Extracurricular), where a score of 1 = high involvement and 3 = low involvement. This might be a model used to predict admission to a university, where it is assumed that high grades, high test scores, and high involvement in activities represent the kind of well-rounded people an admissions office would like to recruit to its university.

The first step is to decide how you want to enter the independent variables into the regression analysis. One method is to list them in the order you want them to be entered, based on some theory or model. The order can make a difference: Given that high school grades and standardized test scores are likely to be correlated, if you enter high school grades first, then test scores will contribute less to the overall final prediction results. On the other hand, if you enter test scores first, then grades might get suppressed. Or you can let the SPSS computer program (or other statistics program) assist you by selecting a stepwise method for entering the variables.

With stepwise, the program calculates the unique contributions of the independent variables to the dependent variable and lists them according to their order of strength. Those that are not statistically significant, based on a t-test and alpha level of .05, are not entered. Think of it as requiring an ID to get into the party: If you are not of "strength," you have to stay outside!

Stepwise multiple regression is selected in Tables 9.4a to 9.4d, and the results are reported in several ways. Although more advanced statistics books and websites go into details about all the components of a regression equation and output, for our purposes, the SPSS information most needed to interpret them is discussed here. The section numbered Table 9.4a is the overall summary of the regression model that has been calculated. Because it is stepwise, it presents the model in steps. Model 1 shows "HS GPA" entered first (according to footnote a). Its correlation (R) with the dependent variable is 0.400, signifying a moderate to strong relationship. This one variable explains 16 percent of the variance in grades, according to R square (R^2). It is also at this point the same as the bivariate Pearson r between high school grades and college grades.

Then the computer program looks for the second strongest predictor or independent variable. Model 2 finds "TOTSAT" and, as footnote b reminds you, at this step, it is high school grades and total SAT scores *combined* that correlate 0.425. R^2 tells you that 18 percent of the variance in CUM GPA in this sample can be explained by respondents' high school grade point averages and total test scores, although including test scores increased variance explained by only 2 percent. This small increase in explanation probably occurred because grades and test scores are themselves related, thereby illustrating the partial effects of a new variable. If high school grades were not included in the list of independent variables, perhaps test scores would have correlated even higher with college GPA.

The computer program keeps calculating and tries a third time, and model 3 shows the arrival of "Extracurricular" to the equation. Multiple R increases to 0.440, and over 19 percent of the explanation of college grades can be accounted for by these three variables together.

Table 9.4b tests the significance of the models that have been generated. Using ANOVA, an F-value is calculated for the regression equation at each step. As can be seen, the final model 3 is a statistically significant regression equation ($p < .000$). The footnotes also remind you of the dependent variable and the independent or predictor variables at each step.

By this point, the overall R^2 is known, but you still want to figure out how each of the independent variables is contributing to this R^2 in explaining or predicting the criterion or dependent variable. Each of them is not equally related or predictive of college grades, especially because there is some interrelationship among the

Table 9.4 Example of Multiple Regression, SPSS

a MODEL SUMMARY

Model	R	R Square	Adjusted R Square	Standard Error of the Estimate
1	0.400[a]	0.160	0.159	0.56952
2	0.425[b]	0.180	0.179	0.56282
3	0.440[c]	0.194	0.192	0.55842

[a] Predictors: (Constant), HS GPA.

[b] Predictors: (Constant), HS GPA, TOTSAT.

[c] Predictors: (Constant), HS GPA, TOTSAT, Extracurricular.

b ANOVA[a]

Model		Sum of Squares	df	Mean Square	F	Significance
1	Regression	69.427	1	69.427	214.046	.000[b]
	Residual	364.579	1,124	0.324		
	Total	434.006	1,125			
2	Regression	78.279	2	39.139	123.559	.000[c]
	Residual	355.727	1,123	0.317		
	Total	434.006	1,125			
3	Regression	84.125	3	28.042	89.925	.000[d]
	Residual	349.881	1,122	0.312		
	Total	434.006	1,125			

[a] Dependent variable: CUM GPA.

[b] Predictors: (Constant), HS GPA.

[c] Predictors: (Constant), HS GPA, TOTSAT.

[d] Predictors: (Constant), HS GPA, TOTSAT, Extracurricular.

independent variables themselves (called *multicollinearity*). The increase in R^2 at each step suggests that a good deal of the relationship is carried by the first variable, high school grade point average. Nonetheless, the other variables still contribute something, but they need to be given less value, or carry less weight. Because high school grades, test scores, and extracurricular participation are measured using different units (grades go from 0.0 to 4.0, combined test scores range from 400 to 1,600, and extracurricular is measured on a three-point scale from 1 to 3), the stan-

c   **COEFFICIENTS[a]**

Model		UNSTANDARDIZED COEFFICIENTS		STANDARDIZED COEFFICIENTS		
		b	Standard Error	Beta	t	Significance
1	(Constant)	1.443	0.108		13.420	.000
	HS GPA	0.456	0.031	0.400	14.630	.000
2	(Constant)	0.852	0.154		5.526	.000
	HS GPA	0.426	0.031	0.374	13.595	.000
	TOTSAT	6.042E−04	0.000	0.145	5.286	.000
3	(Constant)	1.410	0.200		7.049	.000
	HS GPA	0.362	0.034	0.317	10.499	.000
	TOTSAT	5.343E−04	0.000	0.128	4.665	.000
	Extracurricular	−0.129	0.030	−0.131	−4.330	.000

[a] Dependent variable: CUM GPA.

d   **EXCLUDED VARIABLES[a]**

Model		Beta in	t	Significance	Partial Correlation	COLLINEARITY STATISTICS Tolerance
1	TOTSAT	0.145[b]	5.286	.000	0.156	0.967
	Extracurricular	−0.151[b]	−4.991	.000	−0.147	0.796
2	Extracurricular	−0.131[c]	−4.330	.000	−0.128	0.780

[a] Dependent variable: CUM GPA.

[b] Predictors in the Model: (Constant), HS GPA.

[c] Predictors in the Model: (Constant), HS GPA, TOTSAT.

dardized Beta coefficients are used to compare the influence of each variable instead of *b*, the unstandardized ones.

Table 9.4c provides the Beta coefficients for all the independent variables entered at each step in the creation of the model, along with a t-test value and its significance level that tests whether the Beta coefficient is different from zero. As the table shows, at step or model 3, high school GPA has a coefficient of 0.317, total SAT scores have 0.128, and extracurricular has a Beta weight of −0.131. Similar to other correlation coefficients, the negative sign tells you that college grades and participation in extracurricular activities are inversely associated; in this case, those who score low on the extracurricular measure (that is, who selected 1 on the questionnaire where 1 = highly involved in activities) are more likely to score higher on the college GPA measure. However, here it is an inverse association because of the scoring; if 3 was originally written to be highly involved in extracurricular activities, then the Beta weight would have been 0.131. The interpretation is that those highly involved in extracurricular activities tend to have higher GPAs. High school grades and test scores have positive relationships with college GPA.

In addition, Table 9.4d also shows you which variables have been excluded at each step and what their Beta weights would be *if* they are entered in the next step. Notice how at step one under Excluded Variables, the "Beta In" for TOTSAT is 0.145, and sure enough, at model 2 of the coefficients table (Table 9.4c), it enters with a Beta weight of 0.145. Should there be some that never enter the models, as happens with stepwise technique, these would be listed at the last step under "Excluded Variables" with the Beta coefficients they would have had if they had been entered into the model. Sometimes this information can tell you which variables may have been better predictors but were kept out of the model due to intercorrelation with the other variables.

Putting all the findings into words, this regression analysis tells you that around 19 percent of the variation in college cumulative GPA among students in this sample can be explained by knowing their high school grades, SAT scores, and extracurricular involvement. Those in the sample who have higher grades in high school, have higher SAT scores, and are more involved in extracurricular activities tend to have higher college grade point averages. If SAT scores had a negative or inverse coefficient (which they don't), you would conclude that lower SAT scores are associated with higher college grades, and higher SATs are related to lower grades.

The actual equation for the regression line uses the *b* weights and looks like this: $Y' = 1.410 + (0.362 \times X_1) + (0.000534 \times X_2) - (0.129 \times X_3)$, where Y' is the predicted college GPA (dependent variable), X_1 is high school GPA (independent variable 1), X_2 is total SAT scores (independent variable 2), and X_3 is the extracurricular involvement score (independent variable 3). Knowing nothing else, your predicted college GPA would be the value of the constant, 1.410, but your actual prediction is adjusted because you do have some information on the independent variables. High school GPA is giving you stronger assistance in predicting college GPA, compared to both total SAT scores and extracurricular participation, but all three together help out in explaining something about why people get different grades in college. Assuming you are similar to the characteristics of the sample completing the questionnaires that led to the creation of this regression model, you could insert your high school grades, total SAT scores, and estimated participation in high school activities (on a scale of 1 to 3 where 1 = high) and calculate a predicted college grade point average for yourself.

Normally, you don't use a regression equation to calculate information about specific individuals, although many universities use such a model for admissions to predict potential success in college for individual applicants, and many businesses perform regression analyses to estimate future sales about a specific product. Regression analysis does not tell you about any one particular respondent, since the statistics are based on aggregated data. Mostly what you do with regression is construct a profile of characteristics related to the dependent variable from past data and use that to explain what already exists or to predict subsequent outcomes.

The regression does not tell you why any of the findings are so. You can't tell from this why participation in extracurricular activities improves your chances for higher grades in college. Unless you have other data from the questionnaire that can address this query, all you can do when you write up your results is offer some possible explanations based on theory or previous research. Or you can give some speculative interpretations and suggest that maybe those who are more involved in nonacademic pursuits are well-rounded people who exhibit commitment to, interest in, and curiosity about a wide range of issues. Perhaps curiosity and involvement translate into good study habits leading to better grades. Further research would investigate these new hypotheses and research questions, thus illustrating the continuous cycle of inductive and deductive social science research, as discussed in Chapter 1.

There are a variety of other widely used multivariate techniques, such as path analysis, factor analysis, MANOVA, and logistic regression analysis, that are beyond the scope of this introductory book. But the underlying concept is similar: How do two or more independent variables work together in assisting you in making sense of the variation that exists in the dependent variable? How can we account for differences in the dependent variable, knowing two or more independent variables? Similar to other statistical procedures, values are calculated and the probability of obtaining those values by chance is reported. If the odds are fairly small that chance played a role, then you are more confident in declaring results that deserve a big announcement at a press conference or at least a Facebook status posting!

The research journey nears an end. You have completed data analysis and are now ready to put your findings into words and tell others what you have discovered. How to interpret your findings and write a report is the focus of the last chapter.

BOX 9.2
MULTIPLE REGRESSION ANALYSIS IN AN ACADEMIC ARTICLE

Going from elementary to middle school can be a difficult time for many adolescents. Kingery et al. (2011) studied 365 fifth graders in the spring (Time 1) and again in the fall (Time 2) when they were in sixth grade. Let's take a look at some results from what is a much larger study. The researchers asked the students at Time 1 to indicate their "acceptance" of their classmates by having them rate each of their fellow students on "How much do you like to spend time with this person at school?" (where 1 = I don't like to and 5 = I like to a lot). They were also asked to circle the names of friends from the class roster and then answer 40 questions about the quality of a mutual friendship, such as "_____ makes me feel good about my ideas" (where 1 = not at all true and 5 = really true). Here are the results for loneliness, measured with a 24-item Loneliness and Social Dissatisfaction Questionnaire, which asked questions like "There are no other kids I can go to when I need help at school" (where 1 = that's not true at all about me and 5 = that's always true about me).

	LONELINESS		
Predictor variables	β	Standard error	b
Gender[a]	−0.06	0.06	−0.05
Acceptance	−0.12	0.05	−0.20**
Number of friends	−0.05	0.02	−0.18**
Friendship quality	−0.08	0.04	−0.11*
$R^2 = 0.11$			

[a] 1 = female; 0 = male.

n = 339

* $p < .05$, ** $p < .01$

The dependent variable is "loneliness," which is measured using a Likert-type scale, thereby making it suitable for inclusion in a regression analysis. One goal is to predict which students experience loneliness, so the researchers selected four predictor variables, three of which are interval/ratio measures with gender as a dichotomy/dummy variable. As the authors show, the findings presented here are consistent with other studies; that is, loneliness is negatively related to peer acceptance, number of friends, and the quality of those friendships. Gender is not significantly related to loneliness in this sample.

Looking at the results statistically, we can see that "acceptance" has the highest b coefficient at −0.20, followed closely by "number of friends" with −0.18. Given the negative sign, we can conclude that lower rates of acceptance, fewer friends, and poorer quality of friendship correlate with higher degrees of loneliness, as you would expect. Together, all these predictor (independent) variables explain 11 percent (R^2) of the variation of the loneliness scores for this sample of young adolescents. Clearly there are other reasons for feeling lonely in the transition to middle school that make up the unexplained 89 percent.

REVIEW: WHAT DO THESE KEY TERMS MEAN?

Antecedent
Beta coefficients
Dummy variables
Elaboration
Interactive effects
Intervening
Multicollinearity

Multiple correlation
Multiple regression
Partial correlation
Partialling
Regression line
R square
Specification

Spurious
Stepwise
Suppression
Two-way ANOVA
Zero-order

TEST YOURSELF

Here are the results of a multiple regression analysis predicting which people have the happiest outlook on life (where 1 = very happy, 2 = pretty happy, and 3 = not too happy). Variables used include sex (1 = male and 2 = female), age (18 to 90), and years of school completed (number of years from 0 to 20).

MODEL SUMMARY

Model	R	R Square	Adjusted R Square	Standard Error of the Estimate
1	0.161[a]	0.026	0.023	0.606

[a] Predictors: (Constant), respondent's sex, age of respondent, highest year of school completed.

COEFFICIENTS[a]

Model		UNSTANDARDIZED COEFFICIENTS		STANDARDIZED COEFFICIENTS		
		b	Standard Error	Beta	t	Significance
1	(Constant)	2.255	0.111		20.370	.000
	Age of respondent	−0.002	0.001	−0.067	−2.394	.017
	Highest year of school completed	−0.024	0.007	−0.114	−3.494	.000
	Respondent's sex	0.045	0.033	0.036	1.359	.174

[a] Dependent variable: General happiness.

1. List the significant independent variables in order of strength. Do not include any that are statistically insignificant.

2. Interpret R and R^2 for a statistical audience.

3. Using the Beta coefficients, put into words (for a general audience) a profile of those respondents who tend to be the happiest.

INTERPRET: WHAT DO THESE REAL EXAMPLES TELL US?

1. Riggle et al. (2010) studied same-sex legal relationships in order to see if previous research suggesting that married adults experience less psychological stress and higher levels of well-being also applied to lesbian and gay relationships. Comparing those lesbians and gay men in committed relationships with those in legally recognized same-sex relationships (domestic partnerships, civil unions, and civil marriages), the researchers reported the following results for a measure of stress:

	STRESS		
Independent Variable	b	Standard Error	β
Sex	0.481	0.189	0.077*
Education	0.286	0.083	−0.104**
Parent	0.092	0.198	0.014
Relationship length	0.038	0.12	−0.098**
Relationship status	0.675	0.215	−0.095*
Adjusted R^2 = 0.046			

Note: The following categories were used: sex: 0 = male, 1 = female; education: 1 = less than high school degree, 5 = PhD or professional degree; parent: 0 = do not have children, 1 = have children; relationship status: 0 = in a committed relationship, 1 = in a legally recognized relationship. Relationship length was measured in years. Stress is measured on a five-point Likert-type scale where lower scores mean less stress experienced in the previous month.

* $p < .01$, ** $p < .001$

   a. State the hypothesis being tested. Why is regression appropriate to use?
   b. What do the Beta coefficients and significance levels tell you?
   c. Explain R^2 and how it can be interpreted.
   d. Put into words what the study has discovered.

2. Consider the hypothesis that there is a relationship between sex of respondents and whether or not they graduate from college. Table 9.5 shows a cross-tabulation of sex and graduation for a sample of 555 students at a university.

   a. Put into words what the table says about the relationship between graduation rates and sex.

b. What do the chi-square and significance level indicate?
   Yet, you feel there may be something else happening to explain graduation rates and wonder if introducing a control variable could elaborate your findings. Because other studies have shown that high school grades are a good predictor of college success, you control for them. Table 9.6 shows the results.
c. Explain what each subtable says in words.
d. What do the chi-square statistics tell you?
e. What do you conclude about the control variable and the original relationship?

Table 9.5   Cross-Tabulation

GRADUATE BY SEX CROSS-TABULATION

			SEX		
			F	M	Total
Graduate	No	Count	103	101	204
		% within sex	32.2%	43.0%	36.8%
	Yes	Count	217	134	351
		% within sex	67.8%	57.0%	63.2%
Total		Count	320	235	555
		% within sex	100.0%	100.0%	100.0%

Chi-square $= 6.79$, $df = 1$, $p = .009$

CONSULT: WHAT COULD BE DONE?

CNN calls you to be on one of its news shows to discuss election results. The interviewer is planning to ask you about what seems to be a difference in the income of people who voted Democratic, Republican, or for other parties. You are the expert and are going to be asked if this is so.

1. How can you be sure that income is related to political party choice? Describe the steps you would take to verify that relationship. What other variables might be relevant?
2. You will also be asked for whom people might vote the next time around. Just focusing on Democratic or Republican, what variables would you use, and what would you do statistically to predict voting choice?

Table 9.6   Cross-Tabulation of Graduate by Sex, Controlling for Grades

HS Grades				SEX F	M	Total
B or lower	Graduate	No	Count	43	59	102
			% within sex	40.2%	45.4%	43.0%
		Yes	Count	64	71	135
			% within sex	59.8%	54.6%	57.0%
	Total		Count	107	130	237
			% within sex	100.0%	100.0%	100.0%
B+ or higher	Graduate	No	Count	60	42	102
			% within sex	28.2%	40.0%	32.1%
		Yes	Count	153	63	216
			% within sex	71.8%	60.0%	67.9%
	Total		Count	213	105	318
			% within sex	100.0%	100.0%	100.0%

For the high school grades B or lower subtable: chi-square $= 0.647$, $df = 1$, $p = .421$

For the high school grades B+ or higher subtable: chi-square $= 4.518$, $df = 1$, $p = .034$

DECIDE: WHAT DO YOU DO NEXT?

For your study on how people develop and maintain friendships, as well as the differences and similarities among diverse people, respond to the following items:

1. Imagine you have found some relationship between race/ethnicity and number of friends. Suggest several test or control variables you could use to elaborate on the original relationship. Be sure to include at least one antecedent and one intervening variable.
2. Look over your hypotheses and develop one to make it more suitable for a multiple regression. Which variables would you include? How would you recode any variables for a regression?
3. If you have actual data, run a regression, and interpret the results in words. Also, illustrate the process of elaboration, and show how you control for a third variable with a series of crosstabs.

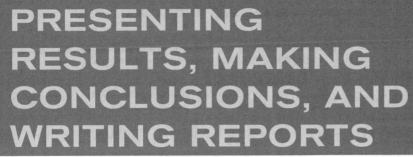

PRESENTING RESULTS, MAKING CONCLUSIONS, AND WRITING REPORTS

10

> Communication is truth; communication is happiness. To share is our duty.
>
> —*Virginia Woolf, writer*

LEARNING GOALS

In this final chapter, learning to write a report of the research project is emphasized, along with the key elements that go into a presentation of your study. Understanding the different audiences reading a report guides the preparation of the findings. By the end of the chapter, you should know the different styles for presenting your research and be able to put together a clear, concise report targeted to the relevant audience.

Y ou've made it this far, and now it's time to show off what you have accomplished. It all began with some idea, hunch, book, or article you read, or maybe even an assignment at work or school. You managed to think up some fascinating research questions and hypotheses, operationalize your variables, and construct a reliable and valid questionnaire. Especially worthy was the sampling strategy you designed to generate representative respondents. You also demonstrated your skill in making decisions about levels of measurement and choosing the most relevant statistics to evaluate the research questions. After all this work, it's time to share it with others and call a press conference to announce your major findings! But first, some important considerations must be kept in mind.

INTERPRETATIONS AND CONCLUSIONS

One of the key ideas to understand in doing survey research is that, although you uncover information with the best scientific methodologies you can employ, it is only one way of apprehending the world around you. Just because you generated a random sample, used valid and reliable measurements, and incorporated the correct statistics does not guarantee in any way that what you discovered is therefore "the truth." However, you have at least used a more scientific way of achieving this information.

Limitations

It's important to remember that some data have not been uncovered, due to the scientific method you used. Some questions remain unanswered, and some information is out of reach. Perhaps qualitative field methods research would have resulted in additional findings, or even contradictory ones, and an experimental laboratory design could have arrived at a more detailed and concise explanation for your outcomes than what you discovered.

In other words, know the *limitations* of your methodology and be sensitive to them when interpreting your results and making conclusions. Most important, know that you have not "proven" anything, merely ruled out alternative explanations or suggested relationships that seem more plausible than others. Knowledge accumulates incrementally, and your contribution is but one step in that process, not the definitive answer to the topic you studied. Hence, avoid such phrases as "my findings prove," "my study conclusively answered all the questions about," or "we found the answer to." It's much more accurate to say that "my findings suggest that," "it seems that in some cases," or "for the sample of people I studied, we can conclude that."

Generalizing Findings

Of utmost importance is the question of *generalizability*. To which other groups or population can you conclude that what you found in your sample also applies? If you studied middle-level managers in a small organization in a medium-size town, can you safely state that the findings are also relevant to middle-level managers in other types of organizations in other kinds of localities? If the results were based on a sample of mostly white respondents, do they also apply to people of other races/ethnicities, or even whites of different educational or social class levels? A frequent problem in research is making conclusions about populations or subgroups that were not sampled appropriately. For most of the research we do, we can safely generalize only about those we actually surveyed or, at best, those similar to the people we studied.

Neutral Statistics?

People often say that statistics don't lie. Well, don't believe that lie (see Darrell Huff's 1954 and still reprinted classic, *How to Lie with Statistics*). Sure, numbers are calculated without bias, but the numbers added and multiplied may have been chosen with intention. Perhaps, for some unethical reason, you decided to ignore the data from certain respondents and not others, or you legitimately eliminated extreme answers and only reported on the middle range of responses. Or maybe you inadvertently led the audience to a particular conclusion using such words as *only* or *very important* or *insignificant* and, in so doing, applied your biased interpretations to the "neutral" statistics.

Imagine you found that 10 percent of the high school sophomores used an illegal drug in the past month. The previous month 8 percent did so. That 2 percent change can also be presented as a 25 percent increase in drug use (25 percent of 8 percent is 2 percent). The headlines might mislead by saying "Drug Use Is 25 Percent Higher Than Last Month," when they should also say that the starting point was 8 percent (see Nardi 2011 for other examples). And what if you said that this month only 10 percent used drugs? The word *only* implies that either you expected more or that 10 percent is not a major issue. On the other hand, antidrug leaders could say that 10 percent is too high and would prefer saying "a disturbing" 10 percent of high school sophomores drank alcohol last month. So much for neutral statistics!

Using Hypotheses to Guide Interpretations

When making conclusions and summarizing findings, keep focused on what your research questions and hypotheses are. Sometimes researchers will write up their results and attempt to explain them using hunches or ideas that were not actually tested in the study. This could be done if the goal is to suggest further research and if the conclusions are written as speculative interpretations. It would not be unusual to say that "perhaps the reason those who studied more get higher grades is connected to the fact that they exhibit more motivation to succeed." But if you haven't included measures of "motivation to succeed," then you cannot say this as if you discovered it in your study. You should phrase interpretations in such a way that those reading your results know the interpretations are speculative, that you were unable to include variables to test them in your project, that you wished you had studied this topic further, and that now you hope someone else will when the research is replicated.

Your summaries should always be in terms of the research questions and literature review you began the project with, and based on the data you actually collected. Certainly, you should not go back and rewrite your hypotheses based on what you have found after you collected the data. Although research does not go as linearly as described in most books like this one, it is not ethical to go back and change what you said you were going to do once you have already done it and arrived at your findings. However, it is good practice to develop new hypotheses and research questions when you uncover some fascinating and serendipitous findings that suggest additional research you can still incorporate in your project. Or perhaps you can adjust measurements and sampling, if you encounter problems along the way with the original design. In all such cases, you must report these modifications in detail in your report.

Cause and Effect Conclusions

Be careful of making conclusions of causality when all you have are measures of correlation without any sense of time sequence or having ruled out other plausible explanations. This is a very common error, as discussed in Chapter 1, even by professional researchers who occasionally imply causation when it was in fact never fully assessed. Often cross-sectional studies are used to suggest longitudinal changes or causation when the evidence is not actually there.

Unless you have data that demonstrate causation, you should conclude with findings focused on relationships and differences, and only speculate or hypothesize about cause and effect. In short, throughout the concluding interpretations, clearly separate speculations from findings rooted in data. Remember to distinguish what you measured for descriptive or exploratory purposes, what you did for explanatory research, and what is being used for predictive purposes. Collecting data to describe relationships is not the same as collecting data for explanatory purposes.

For example, just because you report findings on gender and other information on salaries does not mean there is a causal relationship between them. Let's say you found out that a particular company is 60 percent female and that the average wage is $60,000, compared to another company that is 40 percent female with an average wage of $70,000. You have merely described the organizations in terms of two variables, sex and salary. But to conclude that the percentage of employees that are female is the reason for the lower average wage has not been determined. You must do some additional data analysis that clarifies explanatory processes before you can make such a conclusion.

The Ethics of Reporting Results

In Chapter 2, the topic of ethics focused on the need to be forthright and up front about your project when recruiting subjects. Do no emotional or physical harm, and do allow for informed consent. However, the ethics of doing research don't end with the collection of data.

Data analysis, interpretation, and presentation of results each involve ethical issues as well. For example, if you recall from Chapter 6, when there is a skewed distribution, it's better to use a median than a mean when reporting a central tendency for the variable. Take income: Because few people earn millions of dollars, income is said to have a positive skew, and if the few are included in the calculation of the mean, it could be distorted higher. Imagine a company reporting on the improvement of its employees' salaries by showing an increase in mean wages from one year to the next, only to discover that it was just the executives who saw a dramatic increase in their earnings. If the median were used, the salary figures would probably have stayed about the same, because increasing the chief executive officer's salary from two million dollars to four million keeps the median the same, yet pulls the mean higher. Clearly, this is an unethical use of statistics.

What happens if you get findings that have major repercussions on public policy and people's lives? What are the ethical issues involved in reporting some of your findings and not others, especially those that may run counter to prevailing beliefs or to those of the sponsoring agency of your research? Writing up your report can have a profound impact on people's opinions and behaviors, not to mention their faith in scientific research. How often have you heard that you can't believe what the polls say, yet they rarely have been wrong when done scientifically, as was shown in Chapter 5? How ethical is it for television stations or magazines to report the results of people who logged onto a website survey or called in to vote in a phone poll without offering a disclaimer about the scientific accuracy of the research? What does this do, not only to people's sense of the quality of research but also to their beliefs about the topic voted on?

Interpretations of findings should be informed, intelligent, creative, data based, and ethical. They should not be speculative, selective, biased, and dishonest. Your conclusions should demonstrate a familiarity with the subject, theories, and prior research, as well as provide insights and ideas for future research projects.

AUDIENCES AND REPORTS

An important consideration in the preparation of a report is who will be reading it. This is a basic tip often given by people specializing in teaching good writing. Know your audience. If you visualize to whom you are addressing the remarks, writing ability improves. Any summary of what you have done will always contain some key pieces of information, especially about the methodology, but how it is stated and what elements need to be emphasized might vary, depending on whether your audience is

- The general public
- A reader of an academic journal
- A participant attending a presentation at a conference
- A professor evaluating your thesis
- An employer
- A funding agency

Let's look at how the presentation of research findings might differ depending on the group to whom you are addressing your remarks.

The General Public

When communicating with the general public, avoid academic jargon from your field. There's nothing worse when describing the results of a project than to lapse into concepts that carry very different meanings inside an academic discipline compared with what the public might understand them to mean. For example, someone might describe walking up to people on the street as "randomly" stopping them to ask for their participation in a survey. We, however, know this is a convenience sample, not a random sample. Or imagine reporting the results of a study on "anomie" and "cognitive dissonance" and failing to clarify those terms to a perplexed audience of non–social scientists. Too often we forget that the language we've become accustomed to is impenetrable to those not in our field of expertise.

In addition, most people are not familiar with statistical procedures, so presenting the results of a "multiple regression analysis" and the accompanying "Beta coefficients" will mean nothing to most people. Many of these statistics provide you with the information you need to decide whether your findings are statistically significant or not. For the purposes of a nontechnical audience, the results are the key point, summarized with clarity and insight, not laden with numbers that tell them very little. It is not the time to give a lecture on multivariate statistical analysis during a presentation of your conclusions.

When speaking with the popular media about a study, it is crucial that you clearly describe the limitations. There's nothing worse than reading in some local paper that you "proved" that men are less likely than women to ask for directions when lost, if in fact you only demonstrated a slight tendency to do so in your limited sample of college-age students in a small southeastern university. Frame results with such phrases as "some men demonstrated that" or "it's not clear if these findings apply to older males or those living in other regions of the country."

Readers of an Academic Journal

The best professional publications are those in a particular field that are refereed. This means that every submission is reviewed by selected experts who typically do not know who wrote the article. The author also does not know who is reviewing the paper. This is called a *double-blind* method because both authors and evaluators remain anonymous, thereby allowing for a potentially unbiased review without any preconceived assumptions about the level of expertise or skills of those involved. Reviews for the article are either a "rejection," a "revise and resubmit," or occasionally an outright "accept." A "revise and resubmit" usually is a statement that the paper has some potential but that further clarification, reanalysis of data, rethinking of the interpretations, or other important changes need to be made before another critical review occurs. Then the revised paper is sent back to some or all of the original reviewers, and occasionally to new ones, for another evaluation.

All professional journals (such as the *American Sociological Review*, *British Journal of Sociology*, *American Ethnographer*, and *Journal of Clinical and Social Psychology*, to name a few) provide a "style sheet" about the kind of information expected in the article and the standard format for writing the paper and bibliographic references. These guidelines, often published in issues of the journals or on their websites, also tell you whether these are refereed journals. An academic article is normally organized in the following way (similar to the ordering of the chapters in this book):

- Abstract
- Literature review
- Methods
- Results
- Conclusions and summary
- References

Abstracts. Most articles in the social sciences begin with an abstract, a 100- to 150-word summary of the key points of the research organized in the same order as they are in the complete report or article. Consider it sort of a brief version of an *executive summary*, which is typically a short, one- or two-page summary of the central findings of a study, so termed because it is read by those in leadership positions who supposedly have very little time to read the entire research. An abstract gives readers key ideas about the paper but is not meant to include everything about the research. It should read coherently and stand alone as a self-contained summary of the research with minimal jargon. Abstracts can be descriptive and simply provide the details of the study; these types of abstracts require further reading of the article to find out the detailed results. Other kinds are informative and give the main findings and conclusions. Remember that in this electronic age, many people learn about research through computer databases that provide titles and abstracts. Therefore, the abstract should be clear enough for people outside the field to read and should include key words that would result in the research showing up in an electronic search.

Literature Review. Following the abstract, the body of the paper typically opens with a few sentences stating the focus of the research and its themes. Then a review of the relevant literature serves to frame the paper in some theoretical and empirical context. As described in Chapter 2, a literature review involves a type of content analysis in which you seek out thematic links among the relevant readings and organize the information into those themes. Discuss the key findings of other research, and uncover common threads and differences that run through them. Your goal is to summarize and analyze the research that has been done, raise critical questions about what may be missing from the prior research, and make a case that your research will extend, revise, or replicate what has gone before.

Some people prefer to organize a literature review by summarizing each article or book, one after another. These kinds of literature reviews are similar to annotated bibliographies. However, a literature review should have some analysis of the material and not be just a descriptive listing of research studies with brief summaries of the findings. A review should consider specific themes that emerge from the research that has been done and be organized according to the issues, variables, and theories you are using. How generalizable are the results of the past research? Do the findings apply to the sample your study is using? Are certain variables and measures more valid than others when studying your topic? What theoretical perspectives guided previous studies, and how do they relate to your research goals?

Methods. Most important is the description of the research methods. This should be sufficiently detailed to guide readers in evaluating the quality of the data and research design. The methods section also should provide enough information to serve as a model for those who wish to replicate the study. Include specific information about the sampling procedures (how, when, where, who), composition and size of the final sample (demographics of respondents), the research design and methods used (questionnaire survey, interviews, experimental design, content analysis, participant observations, archival analysis, cross-sectional or longitudinal, etc.), description of the measurements (such as operationalizations, reliability and validity of any standardized instruments), and any other relevant information about the procedures used to gather data (online or distributed questionnaires, number and type of follow-ups to improve response rates, etc.). See Box 10.1 for an example of a methods section write-up.

BOX 10.1
WRITING METHODS

Here's an excerpt from a published academic article explaining how the researchers accomplished their data collection. The information provided illustrates what is needed to assist readers in evaluating the quality of the data and the overall research project. Detailed methods sections also help those interested in replicating a study. The goal for writing a methods section of a report or article is to be as specific as possible with anything that was done that could affect the quality of the data, the interpretation of the results, and the generalizability of the findings.

Schuster et al. (1998: 222–223) report the results from a questionnaire survey focused on sexual behaviors and sexual risk among high school students in a Los Angeles County school district. Notice how detailed the researchers are in describing everything from the ethical issues to the exact procedures for responding and sealing the questionnaire in an envelope. The article also presents information about the subjects, possible absentee rates, and items on the survey.

The district has two public high schools: a general school with about 2,500 students and an alternative school with about 125 students considered at high risk for dropping out. Students in English as a Second Language classes and in intensive special education classes at the general school did not take the survey.

About 12 percent of students were absent from class on the day the survey was conducted at the general high school; the absentee rate at the alternative school was not available but is typically about 35 percent. Of 2,066 students present in appropriate classes on survey day, 2,026 (98 percent) completed the survey, thirty-five did not (they or their parents declined participation), and five turned in unusable surveys....

The survey covered demographics; sexual beliefs, attitudes, and behaviors; and condom use. Male and female versions were identical except for pronouns and sexual behaviors....

To minimize confusion about types of sexual behaviors, we used both precise technical language and anatomic descriptions and avoided euphemistic language.... We adapted the anatomic descriptions from a Centers for Disease Control and Prevention survey and the Surgeon General's Report on Acquired Immune Deficiency Syndrome. All terms and concepts were part of the district's ninth-grade health curriculum....

Respondents completed the anonymous self-administered survey during a regular class period and sealed it in an opaque envelope. Survey administrators unaffiliated with the district proctored the classes. RAND's Human Subjects Protection Committee approved consent and administration procedures.

The school district notified parents about the survey and gave them the opportunity to sign a form denying permission for their children to participate. Students could also decline participation, and names of respondents completing the survey were not recorded. Respondents were instructed to skip questions they preferred not to answer.

Results. This section should be an unbiased and neutral presentation of the results in the context of the research questions and hypotheses. Describe the statistical analyses, provide examples illustrating the results, and give quotations from the open-ended questions or interviews, if any. This is where your analytic skills are demonstrated as you correctly apply the best statistics for the kind of data you have. Be sure to present your results in clearly labeled tables, graphs, and other figures with a title describing in a few words what the figure or table represents, how many people responded, what statistics were used, and which ones are significant findings (usually with asterisks). See the examples presented throughout this book.

Conclusions and Summary. The final section is where interpretations go and where you are encouraged to summarize and make sense of the findings. Here you need to clearly state what is based on speculations and what is rooted in the data. References to the theoretical and empirical findings described in the opening literature review section serve as a way of framing your conclusions. It is common to end your paper with some ideas for future research as well as with some caveats as to what went wrong, what the limitations of the study are, what should be interpreted with caution and why, and how far the results can be generalized beyond the sample. Read some of the academic articles listed in the References section of this book for examples of conclusion sections, literature reviews, methodology descriptions, and abstracts.

References. The last part of any paper is a complete list of references. A bibliography of what you have read should be in the academic discipline's approved style or in the format of the publication or conference receiving your submission. Although most citation styles are slight variations of each other, one of the major differences is the use of footnotes or in-text citations. Most social science journals avoid footnotes for citing research (and if there are any, they tend to be endnotes and put at the conclusion of the paper before the references) and instead refer to a publication in the body of the article by listing the author's name, publication date, and page number if there is a direct quote from the source. The in-text citations and the bibliographic information in the references section of the book you are now reading illustrate the American Sociological Association's publication style and format.

To illustrate, when summarizing a theme of Erving Goffman's classic book *The Presentation of Self in Everyday Life*, you would write, As Goffman (1959) said, the idea of a backstage and frontstage is central to social interaction. For a direct quote from the book, you would add the page number: Goffman (1959: 45). In the alphabetical list of references at the end of the paper, put Goffman's name with the year of publication, book title, and other bibliographic information in the format of the journal's style:

Goffman, Erving. 1959. *The Presentation of Self in Everyday Life*. New York: Doubleday.

A Presentation at a Conference

Similar to the readers of an academic journal, conference audiences are skilled in understanding the jargon of the profession and usually its methodologies. Hence, a paper presented at the annual meetings of a professional organization (such as the American Political Science Association, the American Psychological Association, and the Canadian Anthropology Society) typically follows the format of one submitted to an academic journal. The same kinds of information are required in order for it to be reviewed and judged acceptable for a conference.

However, presenting the paper is a different story. It is not usually the most pleasant experience to have a paper read aloud to you while sitting in uncomfortable chairs in an overly air-conditioned or heated hotel meeting room. Given only 15 or 20 minutes to summarize what can be anywhere from 25 to 50 pages long does not allow for the details that should be in a written version for submission to a journal. Most conferences require that the paper not have been published already. This usually means it's a work in progress or very nearly a finished product. In either case, only key features can be presented at a meeting.

Begin with a brief statement describing the key focus and themes of the research. It is not necessary to review the literature in depth other than to say what your theoretical and empirical context is for the study. For example, instead of going through all 20 articles you researched on alcohol use among high school students, you simply state that your project grew out of sociological labeling theory and previous research that found an increase in substance use around the age of 15, or whatever the data have demonstrated. Then move on. Your goal is to share your findings, not the background research you did before you collected your data.

Highlight just enough about the sample and methodology to provide general information to the audience to assist in interpreting the results. There rarely is enough time to go into details about the measures used and their operationalizations. Many of the specifics about the sample and the data collection procedures are best saved for a handout, overhead projection, or a PowerPoint presentation. Similarly, reading numbers is sure to create a glazed look and droopy heads among audience members, so it's best to save the specific findings for handouts and other means of visual presentation.

Highlighting the key findings of the data analysis should be the central task. Invoke your creative skills to describe the numerical findings verbally and clearly. Rather than say that "there is a statistically significant t-test at the .05 level showing a difference in mean scores on a happiness scale where men scored 3.67 and women scored 3.84 (5 is very happy), and age has a Beta coefficient of 0.58 in a multiple regression predicting happiness," simply state that "there is a statistically significant difference in happiness scores between men and women, and age is the strongest predictor of happiness in a multiple regression analysis." Save the numbers for a handout or later publication.

After discussing the main findings, provide some interpretation of the results and end with a few limitations of the research and any suggestions for future research. Leave time for questions and discussion among the presenters and audience. It is crucial to prepare your presentation by rehearsing out loud and with a watch. Ideally, you should not read your paper; instead, talk about it. Imagine a conversation with a friend (or better yet, rehearse by having such a conversation about your research) and include all the pertinent information for your audience to get a clear sense of what you did and what you found.

Do not try to tell everything about your project. Stay focused on the main findings, highlight the methodology but do not get bogged down in its details, avoid extensive summaries of previous research and theories, and raise some provocative questions that the data have suggested. Your objective is to engage the audience in thinking about the topic in a new way, or to hear some fascinating findings. It is not necessary to report everything you found, because you have very limited time. Those details should be made available to anyone who would like to read the extended version, ideally when it is published in an academic journal. Think of your public presentation as an extended version of an "executive summary."

Professor Evaluating Your Thesis

Many undergraduates are expected to complete a "senior thesis" to graduate with honors or as part of the requirements for their major. For those in graduate school, a thesis or dissertation is central to most master's and doctoral degree programs. A thesis is a very detailed report of your research in which you demonstrate mastery of the literature, expertise in designing relevant methodologies, skills with statistical data analysis, and creative insights in interpreting your findings. Not unlike an academic article in a professional journal, most theses follow the same format, often written as chapters. What is specifically needed for your thesis and format should be supplied to you by those who supervise the research or who are in charge of overseeing the guidelines for conducting and reporting research.

After an initial introduction setting out the themes and purposes of the project, there is usually a very extensive literature review. Most reviews of prior research that appear in scholarly journals are abbreviated versions of what might be in a dissertation. Here you are expected to show off your command of the field, demonstrate how extensively you prepared by citing relevant previous studies, and synthesize the broad literature into meaningful interpretive categories and themes. In many theses, this is where you develop some new theoretical perspectives and ideas. It is also after this review that you present your research questions and hypotheses.

Within the limitations of what is likely to be an underfunded project, you next describe the methodology. Like the material required for a journal article, you spell out your sampling strategies, describe the respondents who filled out your questionnaire, explain the measurements and operationalizations, and discuss the procedures you used to collect the data. This section or chapter should provide enough detailed information not only for others to evaluate the results properly but also to allow them to replicate the research design.

Presentation of your results and data analysis are the core of a thesis. This may take one or more chapters, depending on how the data are organized and what the research questions are. The use of tables, graphs, and other figures is an excellent way of showing what you have found. It is also important to remember that even if none of your hypotheses were supported, no findings are often findings in themselves. Sometimes demonstrating that there is no relationship between some variables or that, for example, there are no sex differences in study habits and grades, can be a revelation.

Your conclusions reflect a scholarly interpretation of the data and should demonstrate your expertise in the field. Reference to the extensive literature you reviewed allows you to contextualize the research and the findings in the theories and empirical studies that came before. Here also is the place to mention any shortcomings about the project and ideas for future research on the topic. A bibliographic list of the extensive readings ends the thesis in the format required by the department or professional field. After it is evaluated and approved by your professor and perhaps other readers, you graduate with a sense of accomplishment and evidence of your abilities to carry out scholarly research.

Employer or Workplace Committee

And you thought doing theses to complete a degree was the last time you had to do this kind of research! Many people find themselves in work situations where they are asked to carry out some research for their company. Often this involves evaluation research, where the effectiveness of some program or new policy implementation needs to be investigated. Or perhaps the nonprofit community service agency for which you volunteer seeks help in designing a study to evaluate its programs and the clients' satisfaction with them. You are asked by your employer or some board to present the results of your project.

In many cases, the suggestions described previously also apply here. Presenting this kind of research is often a combination of speaking to the general public and at a professional conference. It is unlikely that you will be asked to go into the depths you normally would for an honors thesis, or with the jargon and theoretical details of an academic article. Yet, you must be able to demonstrate a sophisticated and scientifically sound research design because any related policy decisions can have profound impact on people's lives. In any case, you have to provide at least an executive summary that clearly and succinctly summarizes the key findings of your research and any relevant methodological points. Being able to condense a major project into one or two pages of information is a skill well worth working on when dealing with employers and other agency boards and organizations.

An extensive literature review is usually not presented in such reports, although doing one best serves your own personal interests because it becomes the source of information about what others have already discovered in this area. If you are hired by the university to study attrition and retention rates, it pays to read the numerous studies that were conducted previously. Why reinvent the wheel? See what measures other researchers used, what questions they addressed, and how they approached the topic, and then modify according to your own unique situation. However, it may not be necessary to write up as detailed a summary as you would be expected to do in a thesis or even an academic article. The key is to present the context for your study and the necessary comparison figures so that those hearing the results can properly interpret what you found.

Careful attention to the methodology is crucial because what you report can easily be discredited if there are major problems with the sampling, operationalizations of the measures, and the overall research design. It is necessary, then, to detail what you did when presenting your findings, whether in writing or in an oral presentation, and to explain your decisions.

Results of your data analysis are more than likely being distributed to people who do not have extensive knowledge of the range of statistical procedures employed. Again, some of the statistical analyses may be for your own background information, so that you can say with confidence what the findings are and whether they are statistically significant. Many agencies and workplaces are satisfied with descriptive data, cross-tabulations, and graphs. More complex data analyses, such as multiple regression, ANOVA, and even the simpler statistics like chi-square, Pearson r correlation, and t-tests, rarely get presented to employers and agency boards, unless there is some expectation to do so by those who are qualified to read and interpret the information. Statistics could be placed in an appendix with an occasional reference that statistical analyses of the data were completed and that these verify the significance of the results presented.

Many agencies or employers ask that you also present interpretations and suggestions for policy changes in light of the findings. However, some would rather leave that to the agency or another committee formed to make sense of the results and implement policy changes in the workplace. Again, this depends on the organization, and you should be clear about the expectations when you are assigned the project. So, for example, if you uncover with a multiple regression analysis that the quality of the campus facilities is a more important explanation than the quality of the courses for a rise in attrition rates at the university, then you may be asked to account for this and what you would recommend to change it. Or you might find that the data from an evaluation of a training program for new employees at the organization you studied indicate that they should be compensated for participation in a three-day orientation program. Depending on your role in the project, you might be asked to provide a recommendation, so you suggest that the employer should implement a training compensation program.

Grant or Funding Agency

If you were lucky enough to have your research funded through some grant or funding agency, then the requirements for disseminating the results of your study were probably spelled out when you received the grant. Normally, a report is due by a specified date, and it can vary from a copy of the entire thesis or research project to a long executive summary of 5 to 10 pages.

It is also very likely that, in order to have received the grant, you have already sent the agency a detailed proposal that included a literature review and extensive methodological procedures. Most funding agencies pay particular attention to the methodology and the ethical dimensions of the project for research with human subjects. Typically, a recommendation from the Institutional Review Board (IRB) is required. A budget was probably also requested, and your final report may have to account for how the money was actually spent.

A grant report usually includes a summary of the data analysis, the statistics used, and your interpretations of the results. You are expected to demonstrate how the objectives and other goals you described in the initial grant proposal were or were not achieved. As always, reporting the limitations of the study and providing ideas for future research topics are good ways of concluding your summary to the agency. The format of a report follows the outline of topics and sections required in most academic articles and theses. The exact kinds of information requested are normally discussed in the granting agency's guidelines.

THE JOURNEY FINISHES

Congratulations! You have completed your journey along the research trail. A few final words are in order: It's one thing to read about doing research; it's another to actually do it. Imagine reading about how to drive a car or play the piano and never actually driving or playing. It's the same with research. You cannot read it and think that you now know how to do it. Just like driving a car and playing the piano, you have to practice, practice, practice. You have to get out there and drive, play, and research.

Do not expect to do research exactly the way it sounds in this book. Research involves making trade-offs and compromises. It may seem ideal to get a random sample, but what happens when that generates a low response rate? You compromise and seek out a convenience sample. Sure you would like to have an interval/ratio measure of income, but when many respondents refuse to answer, you realize that you might have to compromise to reduce nonresponses and ask the question using ordinal categories of income instead.

Typically, research lurches forward, is set back for any number of reasons (delays in getting questionnaires completed, typos and other errors only noticed later, funding problems, computer glitches, and other petty annoyances), gets refined when some other unexpected finding emerges in your data, or reinterpreted when another article is discovered that sheds new theoretical light on your results. Your findings may lead to a modification of the theory you started out with in an ongoing deductive and inductive process. Although the report or article you complete presents a summary in a linear fashion and thereby reinforces the illusion, it might make for interesting reading if you included a list of shortcomings and caveats, along with a few of the ways your project veered from the standard journey.

Positivism

Because this book focuses on survey research methods and does not cover such techniques as content analysis, field methods, and other qualitative research methods, remember that writing a questionnaire to gather data carries with it some assumptions about the nature of reality and how we understand it. It is just one way of apprehending social phenomena and not the only way. It has its limitations and it has certain strengths. And it is based on a particular belief about understanding the world around us called positivism.

Positivism is the system of thought developed in the early 1800s in France by Auguste Comte, a philosopher and founder of sociology, that emphasizes empirical data, reasoning, and the use of scientific principles to explain social phenomena. It is a philosophy of knowledge that stresses the experiential, that is, what is experienced through the senses and what is measurable using scientific techniques. Positivism believes reality is something that can be assessed, measured, and discovered logically. Opposed to this view of the world is the more metaphysical and humanistic belief that ideas are personal and cannot be measured scientifically. All human ideas, including the scientific method, are subjective creations. Truth is not something easily measured.

Whichever philosophy of knowledge makes sense to you, at least understand that choosing a questionnaire survey method implies signing on to a philosophy that assumes that the phenomena you wish to study are measurable and open to empirical assessment. Doing survey research can answer some questions, and then only with the limitations imposed by its methods. It is just one way of understanding the opinions and behavior of humans, but not the only one. The use of multiple techniques, what is sometimes called *triangulation*, to understand complex phenomena is ideal. Sometimes a self-report questionnaire should be combined with face-to-face interviews and participant observation of behavior to comprehend more fully the complex behaviors and attitudes you wish to study.

However, in face of the limitations imposed by funding and time, the selection of the most appropriate methodology must be determined by the nature of the phenomenon and population under scrutiny and the research questions you want answered. Clearly, a study involving young children who cannot read or people with cognitive limitations should not depend on survey research that uses questionnaires. Research focused on how people behave in particular situations, such as children at play or employees at work, may best be served with participant observation field methods. It is never ideal to begin by saying you want to write a questionnaire and then choose your topic.

Let the research problem lead you to the best methodology. And if a questionnaire serves the goals of your project, then let this book be your guide for the research journey.

REVIEW: WHAT DO THESE KEY TERMS MEAN?

Abstract	Ethics	Organization of a journal article
Academic journal	Executive summary	Positivism
Audiences	Literature review	Triangulation

TEST YOURSELF

Take this statistical finding: Researchers found a 0.65 Pearson r correlation between scores on a Likert-type social justice scale (where 1 = low belief in social justice and

5 = very strong belief in social justice) and the number of times respondents had volunteered in the previous six months at a homeless shelter or equivalent nonprofit organization.

Indicate how you would write this result up for the following:

1. An academic journal
2. A presentation to a nonprofit organization
3. A report to a city council debating cutting funds to a local social service agency
4. An online blog you write
5. A Twitter statement (limit 140 characters)

INTERPRET: WHAT DO THESE REAL EXAMPLES TELL US?

1. For this excerpt from a published academic article, discuss what is included and what might be missing from the final write-up of the methods section.

   Shelton et al. (2010: 76) designed a study to explore the processes involved in interracial friendship development. Here is what the researchers wrote:

   Participants
   "We recruited 50 white and 24 black students to participate in a study on friendship development for $50 and a chance to win additional monetary prizes in a drawing. The sample consisted of 42 females (14 black and 28 white) and 32 males (10 black and 22 white)."

   Procedures
   "Upon agreeing to participate in the study, all participants attended an orientation session where they were told that they would select two people of their same sex whom they did not know very well at that time but might become good friends with as the semester progressed. We required the participants to select a white and a black potential friend. In addition, we told the participants that they would complete a questionnaire about each of the potential friends every two weeks for the next 10 weeks. We emailed reminders to all participants on the day the questionnaires needed to be completed. The participants completed the questionnaires online or in pencil-and-paper format. We gave participants who completed the questionnaire via hardcopy a campus mail envelope to return the questionnaires as soon as they completed them every two weeks. At the end of the 10-week period, the participants attended a post-study session where they completed a final questionnaire, were informed of the purpose of the study, and received their payment."

2. For a public presentation of ongoing research looking at the effectiveness of Learning Communities in community colleges, Brock (2010) wrote this summary. Just

based on this excerpt taken from a larger report, discuss what is included and what further information you would like to know. How does this presentation summary differ from a more formal academic journal publication?

The Learning Communities targeted incoming freshmen, the great majority of whom required developmental English. Students in Learning Communities were placed into groups of 15–25 that took three courses together: an English course geared toward their level of proficiency; a regular college course like introductory psychology or sociology; and a student success course, taught by a college counselor, that covered effective study habits and other skills necessary to succeed in college. Faculty who taught in the Learning Communities were expected to coordinate assignments and meet periodically to review student progress. The idea was to build social cohesion among students and faculty and to help students apply the concepts and lessons across the courses.

More than 1,500 students participated in the Learning Communities evaluation and were, as noted, randomly assigned to either a program group that participated in Learning Communities or a control group that took regular, unlinked courses. The students were young (mostly 17 to 20 years old), low-income, and highly diverse in terms of race and ethnicity. The research team tracked program and control group members for two years and found that students in the Learning Communities were more likely to feel integrated at school and be engaged in their courses. They also passed more courses and earned more credits during their first semester, moved more quickly through developmental English courses, and were more likely to take and pass an English skills assessment test that was required for graduation. It is important to note that these effects, while statistically significant, were generally modest. For example, after four semesters, students in the program group earned an average of 33.2 college credits, compared with an average of 30.8 credits for the control group (a difference of less than one course). Moreover, contrary to expectations, the Learning Communities did not have an immediate effect on persistence.

CONSULT: WHAT COULD BE DONE?

The local newspaper has heard about a study looking at truancy at the town's high school that has data potentially damaging to the reputation of the teachers and administrators at the school. It also links truancy with substance abuse and violent threats to other students.

1. What kinds of questions would you advise the reporters to ask about the study?
2. What advice would you give the researchers who are about to release their findings?

3. What kinds of information about the study should be included in all reports of it by the study directors as well as the media?

DECIDE: WHAT DO YOU DO NEXT?

For your study on how people develop and maintain friendships, as well as the differences and similarities among diverse people, respond to the following items:

1. In what different ways would you present your findings if you were reporting the results to (a) a school board concerned about friendship issues in the local high school, (b) the board members of a company that sponsored the research, (c) racial and religious organizations interested in bridging differences among diverse groups, (d) an academic journal, (e) the local news media and online blogs, and (f) first-year college students during orientation week?
2. What ethical issues should you be attentive to when presenting the data in each of these situations?

appendix

Statistical Analysis Decision Tree

Determine which variable is *independent* and which is *dependent* in your hypothesis or research question.

↓

Determine each variable's *level of measurement* (nominal, ordinal, interval/ratio). Remember that dichotomies can be treated as any level and many ordinal measures with equal-appearing intervals can be used as interval/ratio measures.

↓

Run *frequencies* and appropriate *descriptive statistics* (such as mode, median, mean) to assess whether each variable is really a variable (and not a constant) in your sample.

↓

Decide on the appropriate statistics to use to *analyze* relationships between the independent and dependent variables.

To *compare means* of a dependent variable (interval/ratio) between *two* categories of an independent variable: use **t-test**. Among *three or more* categories of an independent variable: use **ANOVA**.

To test a relationship between *nominal* and/or *ordinal* variables: use **chi-square**. If at least one variable is nominal: use **Lambda**. If both are ordinal: use **Gamma**. If the variables are ordinal ranks: use **Spearman's rho**.

To test a relationship between two *interval/ratio* variables: use **Pearson r** *correlation*. If there are two or more independent variables: use **multiple correlation (R) and linear regression**.

If the *probability* of obtaining that statistic by chance is less than .05 ($p < .05$), then *reject* the *null* hypothesis of no difference or no relationship, and declare there *is* a *significant* relationship between the two variables.

For Lambda, Gamma, Spearman, and Pearson r correlations, assess the *strength* of the relationship: around 0 to .25 is low, .26 to .60 is moderate, and .61 to 1.0 is strong.
The *direction* (+ or -) also tells you if it's a positive relationship (both increase or decrease in same direction) or an inverse one (as one variable increases, the other decreases).

ANSWERS FOR "TEST YOURSELF" EXERCISES

Chapter 1

1. Before you can conclude cause and effect, you must determine that (a) there is a significant relationship between the variables, (b) the cause (independent variable) came before the effect (dependent variable) in time, and (c) alternative explanations or causes have been ruled out.

2. The article provides several examples, but begin by asking whether there are any significant statistical relationships between autism and vaccines in the scientific literature. If so, did autism show up before or after the vaccines? If it did, then ask what other possible causes could be at work, such as nutrition, urban pollution, genetics, family history of illnesses, and so forth.

3. Common everyday thinking at work in these situations is anecdotal evidence. People hear stories from other concerned parents and see media reports about autism, especially on numerous blogs and other websites. Anxieties about their children often lead parents to seek any explanation that brings them some comfort, especially when ambiguity is present. Scientific studies would require replication among different populations, controls for alternative explanations, and establishment of a timeline of causes and effects.

Chapter 2

1. Because individual names can be linked to code numbers, this is no longer an anonymous survey. It's up to the researchers then to maintain confidentiality and ethically not to disclose specific answers for individual respondents.

2. These guiding principles—autonomy, beneficence, and justice—are defined at the end of the chapter. Try to apply these to an actual research topic, like studying drug use on campus.

3. Discuss how imagination can apply to generating new ideas, like serendipity. Or consider how, when reading other research findings, you creatively come up with an idea based on those findings. Use your imagination, in other words, to answer this question too!

Chapter 3

1. There is no relationship between education level (1 = high school graduate, 2 = some college, 3 = college graduate, 4 = graduate school) and scores on a scale measuring life satisfaction (scores range from 1 to 10, where 10 = highly satisfied with one's life).

	Which Variable?	Level of Measurement?	Type of Hypothesis?
Independent variable	Educational level	Ordinal	Null, two-directional
Dependent variable	Life satisfaction	Interval/ratio	

2. Men are more likely to receive higher hourly wages than women.

	Which Variable?	Level of Measurement?	Type of Hypothesis?
Independent variable	Sex	Nominal (dichotomy)	One-directional
Dependent variable	Hourly wages	Interval/ratio	

3. There is a relationship between ethnicity/race and political party affiliation.

	Which Variable?	Level of Measurement?	Type of Hypothesis?
Independent variable	Ethnicity/race	Nominal	Two-directional
Dependent variable	Political party affiliation	Nominal	

Chapter 4

1. Responses are not mutually exclusive ("5 times" appears twice) and not exhaustive (there's no option for "none" or "more than 7 times").

2. This is a double-barreled question. What if you don't watch sitcoms but only watch dramas? "And" indicates two questions in one.

3. There is no branching. When respondents answer no, they should be told to skip question b.

4. Yes and no responses don't capture the intensity of opinion (perhaps options should range from "strongly agree" to "strongly disagree"), and invoking a scientific authority indicates a possible loaded question, resulting in agreement.

Chapter 5

1. Quota nonprobability.
2. Convenience nonprobability.
3. Systematic random.
4. Snowball.

Chapter 6

1.

	Statistic	Graph/Chart
a. SAT Scores	Mean	Histogram or frequency curve
b. Race/ethnicity	Mode	Pie or bar
c. Skewed number of hours studied in the past week	Median	Histogram or frequency curve
d. Type of car owned	Mode	Pie or bar

2. a. The mean, median, and mode are approximately the same, suggesting a normal curve.
   b. i. 68 percent.
       ii. Under 63.3 inches.
       iii. 50 percent.
       iv. Two-tailed, so either taller than about 75.3 inches or shorter than 59.3 inches.
       v.   Between 59.3 and 75.3 inches.

Chapter 7

1. There is no relationship between SAT scores and college GPA.

	Which Variable?	Level of Measurement?	Which Statistic to Use?
Independent variable	SAT scores	Interval/ratio	Pearson r
Dependent variable	College GPA	Interval/ratio	

2. There is no relationship between type of car owned and region of the country.

	Which Variable?	Level of Measurement?	Which Statistic to Use?
Independent variable	Region of country	Nominal	Lambda, chi-square
Dependent variable	Type of car owned	Nominal	

3. There is no relationship between the Top 20 college football rankings this year compared to these schools' rankings last year.

	Which Variable?	Level of Measurement?	Which Statistic to Use?
Independent variable	Last year's rankings	Ordinal	Spearman's rho
Dependent variable	This year's rankings	Ordinal	

4. There is no relationship between gender and number of times using Twitter per day (measured as "none," "1 to 5 times," "6 to 10 times," "11 or more times").

	Which Variable?	Level of Measurement?	Which Statistic to Use?
Independent variable	Gender	Nominal, but dichotomy so can be interpreted as ordinal	Gamma, chi-square
Dependent variable	Times using Twitter	Ordinal	

Chapter 8

1. a. F-test ANOVA.
   b. Paired samples t-test.
   c. Independent samples t-test.
2. a. Because the survey is comparing average attitude scores on an equal-appearing interval Likert scale between two independent groups (college versus no college).
   b. The Levene test tells us whether or not the variances for each group are equal. In this case, F is significant, so reject the null of no difference in variances and conclude that equal variances are not assumed. The variances are different for each group.
   c. Therefore, you look at the t-value for "equal variances not assumed," that is, 12.205.
   d. Given a t-value of 12.205, with almost 710 degrees of freedom, we see the significance level is $p < .001$, so we reject the null hypothesis and conclude there is a difference in mean attitudes toward classical music between those who have a college degree and those who do not have a college degree.
   e. College grads are more likely to like classical music (with an average of 2.06 where 1 = like very much).

Chapter 9

1. Significant independent variables in order of strength are highest year of school completed, then age of respondent (respondent's sex is not statistically significant).

2. The multiple correlation of the three independent variables together (including the nonsignificant one) with the dependent variable is 0.161. It is a weak correlation. R^2 tells us the PRE: 2.6 percent of the variation in respondents' general happiness can be explained by the independent variables.

3. Those who are closer to 1 on the happiness scale (that is, those who are happier) tend to be older and have more education (the negative Beta coefficients indicate low scores on the happiness scale). The happiest have high scores on the age and education measures; that is, they are older and have more years of schooling.

Chapter 10

There is no one right answer for this one, yet be sure to consider the following items for each audience:

1. Include what the 0.65 indicates, such as very strong correlation, and its R^2 interpretation (42 percent of reasons why they volunteer relate to beliefs in social justice). Explain why a Pearson r was used (two interval/ratio measures).

2. Focus on the connection between attitudes and action, between what people say they believe and how that translates into behavior.

3. Argue for the importance of these shelters and how there is a strong relationship to values of social justice, thereby enhancing the city council's values.

4. Here you can express your personal opinions about social justice and even focus on why there isn't a stronger correlation between what people believe and how they carry out their beliefs.

5. As research suggests, putting social justice into action can mean volunteering at a homeless shelter near you. So do something today.

references

Babbie, Earl. 2010. *The Practice of Social Research.* Twelfth edition. Belmont, CA: Wadsworth.

Baumann, Shyon. 2001. "Intellectualization and Art World Development: Film in the United States." *American Sociological Review* 66:3, 404–426.

Berg, Bruce L., and Howard Lune. 2011. *Qualitative Research Methods for the Social Sciences.* Eighth Edition. Boston: Allyn & Bacon.

Blake, Susan M., Rebecca Ledsky, Thomas Lehman, Carol Goodenow, Richard Sawyer, and Tim Hack. 2001. "Preventing Sexual Risk Behaviors among Gay, Lesbian, and Bisexual Adolescents: The Benefits of Gay-Sensitive HIV Instruction in Schools." *American Journal of Public Health* 91:6, 940–946.

Brock, Thomas. 2010. "Evaluating Programs for Community College Students: How Do We Know What Works?" MDRC. October. Retrieved January 2, 2013. www.mdrc.org/sites/default/files/paper.pdf

Budiansky, Stephen. 1995. "Consulting the Oracle: Everyone Loves Polls. But Can You Trust Them?" *U.S. News & World Report,* December 4, 52–55, 58.

Campbell, Donald, and Julian Stanley. 1963. *Experimental and Quasi-Experimental Designs for Research.* Chicago: Rand McNally.

Catania, Joseph A., Dennis Osmond, Ronald D. Stall, Lance Pollack, Jay P. Paul, Sally Blower, Diane Binson, Jesse A. Canchola, Thomas C. Mills, Lawrence Fisher, Kyung-Hee Choi, Travis Porco, Charles Turner, Johnny Blair, Jeffrey Henne, Larry L. Bye, and Thomas J. Coates. 2001. "The Continuing HIV Epidemic among Men Who Have Sex with Men." *American Journal of Public Health* 91:6, 907–914.

Christian, Leah, Scott Keeter, Kristen Purcell, and Aaron Smith. 2010. "Assessing the Cell Phone Challenge." Pew Research Center. Retrieved January 2, 2013. www.pewresearch.org/2010/05/20/assessing-the-cell-phone-challenge.

Cresswell, John, and Vicki Plano Clark. 2011. *Designing and Conducting Mixed Methods Research.* Thousand Oaks, CA: Sage.

Cyders, Melissa A., Kate Flory, Sarah Rainer, and Gregory T. Smith. 2009. "The Role of Personality Dispositions to Risky Behavior in Predicting First Year College Drinking." *Addiction* 104:2, 193–202.

Diaz, Raphael M., George Ayala, Edward Bein, Jeff Henne, and Barbara V. Marin. 2001. "The Impact of Homophobia, Poverty, and Racism on the Mental Health of Gay and Bisexual Latino Men: Findings from 3 US Cities." *American Journal of Public Health* 91:6, 927–932.

Dillman, Don A. 2007. *Mail and Internet Surveys: The Tailored Design Method.* Second edition. New York: Wiley.

Durkheim, Emile. 1951 [1897]. *Suicide: A Study in Sociology.* New York: Free Press.

Dybbro, Tommy. 1972. "Population Studies on the White Stork *Ciconia ciconia* in Denmark." *Ornis Scandinavica* 3, 91–97.

Esterberg, Kristin. 2001. *Qualitative Methods in Social Research.* New York: McGraw-Hill.

Felmlee, Diane, and Anna Muraco. 2009. "Gender and Friendship Norms among Older Adults." *Research on Aging* 31:3, 318–344.

Flowers, Lamont, Brian K. Bridges, and James L. Moore. 2011. "Concurrent Validity of the Learning and Study Strategies Inventory (LASSI): A Study of African American Precollege Students." *Journal of Black Studies* 20:10, 1–15.

Frankel, Mark S., and Sanyin Siang. 1999. "Ethical and Legal Aspects of Human Subjects Research on the Internet." American Association for the Advancement of Science. Retrieved January 2, 2013. www.aaas.org/spp/sfrl/projects/intres/report.pdf.

Gallup Polls. 2013. "Election Polls—Accuracy Record in Presidential Elections." Retrieved January 2, 2013. www.gallup.com/poll/9442/election-polls-accuracy-record-presidential-elections.aspx.

Goffman, Erving. 1959. *The Presentation of Self in Everyday Life.* New York: Doubleday.

Hansen, Wendy L., and Neil J. Mitchell. 2000. "Disaggregating and Explaining Corporate Political Activity: Domestic and Foreign Corporations in National Politics." *American Political Science Review* 94:4, 891–903.

Hechtkopf, Kevin. 2010. "Support for Gays in the Military Depends on the Question." CBS News. Retrieved January 2, 2013. www.cbsnews.com/8301-503544_162-6198284-503544.html.

Huff, Darrell. 1954 [1993]. *How to Lie with Statistics.* New York: Norton.

Huhman, Marian, Lance D. Potter, Mary Jo Nolin, Andrea Piesse, David R. Judkins, Stephen W. Banspach, and Faye L. Wong. 2010. "The Influence of the VERB Campaign on Children's Physical Activity in 2002 to 2006." *American Journal of Public Health* 100:4, 638–645.

Jones, Rachel K., and Ye Luo. 1999. "The Culture of Poverty and African-American Culture." *Sociological Perspectives* 42:3, 439–458.

Kibria, Nazli. 2000. "Race, Ethnic Options, and Ethnic Binds: Identity Negotiations of Second-Generation Chinese and Korean Americans." *Sociological Perspectives* 43:1, 77–95.

Kingery, Julie, Cynthia A. Erdley, and Katherine C. Marshall. 2011. "Peer Acceptance and Friendship as Predictors of Early Adolescents' Adjustment across the Middle School Transition." *Merrill-Palmer Quarterly* 57:3, 215–243.

Kort-Butler, Lisa A., and Kelley J. Sittner Hartshorn. 2011. "Watching the Detectives: Crime Programming, Fear of Crime, and Attitudes about the Criminal Justice System." *Sociological Quarterly* 52, 36–55.

Laumann, Edward O., John H. Gagnon, Robert T. Michael, and Stuart Michaels. 1994. *The Social Organization of Sexuality: Sexual Practices in the United States.* Chicago: University of Chicago Press.

LeBlanc, Allen J., and Richard G. Wight. 2000. "Reciprocity and Depression in AIDS Caregiving." *Sociological Perspectives* 43:4, 631–649.

McMahon, Susan D., Jamie Wernsman, and Dale Rose. 2009. "The Relation of Classroom Environment and School Belonging to Academic Self-Efficacy among Urban Fourth- and Fifth-Grade Students." *Elementary School Journal* 109:3, 267–281.

Mears, Daniel P. 2001. "The Immigration-Crime Nexus: Toward an Analytic Framework for Assessing and Guiding Theory, Research, and Policy." *Sociological Perspectives* 44:1, 1–19.

Menand, Louis. 2004. "The Unpolitical Animal." *New Yorker,* August 30, 92–96.

Meyer, Hans, Doreen Marchionni, and Esther Thorson. 2010. "The Journalist behind the News: Credibility of Straight, Collaborative, Opinionated, and Blogged 'News.'" *American Behavioral Scientist* 54:2, 100–119.

Meyer, Ilan H. 1995. "Minority Stress and Mental Health in Gay Men." *Journal of Health and Social Behavior* 36, 38–56.

Migliaccio, Todd. 2009. "Men's Friendships: Performances of Masculinity." *Journal of Men's Studies* 17:3, 226–241.

Miller, Delbert C., and Neil Salkind. 2002. *Handbook of Research Design and Social Measurement.* Sixth edition. Newbury Park, CA: Sage.

Mutz, Diana C., and Paul S. Martin. 2001. "Facilitating Communication across Lines of Political Difference: The Role of Mass Media." *American Political Science Review* 95:1, 97–114.

Nagasawa, Richard, Zhenchao Qian, and Paul Wong. 2000. "Social Control Theory as a Theory of Conformity: The Case of Asian/Pacific Drug and Alcohol Nonuse." *Sociological Perspectives* 43:4, 581–603.

Nardi, Peter M. 2011. "Understanding Popular Uses of Percentages." *Pacific Standard.* April 30. Retrieved January 2, 2013. www.psmag.com/culture-society/understanding-popular-uses-of-percentages-30686.

Park, Madison. 2010. "Medical Journal Retracts Study Linking Autism to Vaccine." CNN. February 20. Retrieved January 2, 2013. www.cnn.com/2010/HEALTH/02/02/lancet.retraction.autism/index.html.

Pew Research Center. 2011a. "Question Order." Retrieved January 2, 2013. http://people-press.org/methodology/questionnaire-design/question-order.

Pew Research Center. 2011b. "Americans and Text Messaging." Retrieved January 2, 2013. www.pewinternet.org/Reports/2011/Cell-Phone-Texting-2011.aspx.

Phillips, Susan, Alan J. Dettlaff, and Melinda J. Baldwin. 2010. "An Exploratory Study of the Range of Implications of Families' Criminal Justice System Involvement in Child Welfare Cases." *Children and Youth Services Review* 32:4, 544–550.

Plummer, Ken. 2001. *Documents of Life 2: An Invitation to a Critical Humanism.* London: Sage.

Reiss, Ira. 1993. "The Future of Sex Research and the Meaning of Science." *Journal of Sex Research* 30:1, 3–11.

Riggle, Ellen, Sharon S. Rostosky, and Sharon G. Horne. 2010. "Psychological Distress, Well-Being, and Legal Recognition in Same-Sex Couple Relationships." *Journal of Family Psychology* 24:1, 82–86.

Rosenberg, Morris. 1968. *The Logic of Survey Analysis.* New York: Basic Books.

Rossi, Peter, Mark Lipsey, and Howard Freeman. 2004. *Evaluation: A Systematic Approach.* Thousand Oaks, CA: Sage.

Schuster, Mark A., Robert M. Bell, Gene A. Nakajima, and David E. Kanouse. 1998. "The Sexual Practices of Asian and Pacific Islander High School Students." *Journal of Adolescent Health* 23, 221–231.

Shaw, Clifford, and Henry McKay. 1942. *Juvenile Delinquency and Urban Areas.* Chicago: University of Chicago Press.

Shelton, J. Nicole, Thomas E. Trail, Tessa V. West, and Hilary B. Bergsieker. 2010. "From Strangers to Friends: The Interpersonal Process Model of Intimacy in Developing Interracial Friendships." *Journal of Social and Personal Relationships* 27:1, 71–90.

Simon, Valerie, and Wyndol Furman. 2010. "Interparental Conflict and Adolescents' Romantic Relationship Conflict." *Journal of Research on Adolescence* 20:1, 188–209.

Slater, Michael D., and Andrew F. Hayes. 2010. "The Influence of Youth Music Television Viewership on Changes in Cigarette Use and Association with Smoking Peers: A Social Identity, Reinforcing Spirals Perspective." *Communication Research* 37:6, 751–773.

Smith, Tom W., Peter Marsden, Michael Hout, and Jibum Kim. 2011. "General Social Surveys, 1972–2010" (machine-readable data file). NORC ed. Chicago: National Opinion Research Center (producer); Storrs, CT: Roper Center for Public Opinion Research, University of Connecticut (distributor).

Southgate, Darby E., and Vincent J. Roscigno. 2009. "The Impact of Music on Childhood and Adolescent Achievement." *Social Science Quarterly* 90:1, 4–21.

Survey Monkey. 2011. "How to Estimate Your Population." Retrieved January 2, 2013. http://blog.surveymonkey.com/blog/2012/10/02/how-to-estimate-your-population.

Sutherland, Edwin. 1934. *Principles of Criminology.* Chicago: Lippincott.

Suzuki, Etsuji, Soshi Takaoa, S. V. Subramanianb, Hirokazu Komatsua, Hiroyuki Doia, and Ichiro Kawachi. 2010. "Does Low Workplace Social Capital Have Detrimental Effect on Workers' Health?" *Social Science and Medicine* 70:9, 1367–1372.

ter Bogt, Tom, Rutger Engels, Sanne Bogers, and Monique Kloosterman. 2010. "'Shake It Baby, Shake It': Media Preferences, Sexual Attitudes and Gender Stereotypes among Adolescents." *Sex Roles* 63, 844–859.

Weber, Robert Philip. 1990. *Basic Content Analysis.* Second edition. Newbury Park, CA: Sage.

Zuo, Jiping, and Shengming Tang. 2000. "Breadwinner Status and Gender Ideologies of Men and Women Regarding Family Roles." *Sociological Perspectives* 43:1, 29–43.

index